Ethical Ripples of Creativity and Innovation

Ethical Ripples of Creativity and Innovation

Seana Moran
Clark University, USA

Editorial matter and selection © Seana Moran 2016
Individual chapters © Seana Moran and respective co-authors 2016

All rights reserved. No reproduction, copy or transmission of this publication may be made without written permission.

No portion of this publication may be reproduced, copied or transmitted save with written permission or in accordance with the provisions of the Copyright, Designs and Patents Act 1988, or under the terms of any licence permitting limited copying issued by the Copyright Licensing Agency, Saffron House, 6–10 Kirby Street, London EC1N 8TS.

Any person who does any unauthorized act in relation to this publication may be liable to criminal prosecution and civil claims for damages.

The authors have asserted their rights to be identified as the authors of this work in accordance with the Copyright, Designs and Patents Act 1988.

First published 2016 by
PALGRAVE MACMILLAN

Palgrave Macmillan in the UK is an imprint of Macmillan Publishers Limited, registered in England, company number 785998, of Houndmills, Basingstoke, Hampshire RG21 6XS.

Palgrave Macmillan in the US is a division of St Martin's Press LLC, 175 Fifth Avenue, New York, NY 10010.

Palgrave Macmillan is the global academic imprint of the above companies and has companies and representatives throughout the world.

Palgrave® and Macmillan® are registered trademarks in the United States, the United Kingdom, Europe and other countries.

ISBN: 978–1–137–50553–8

This book is printed on paper suitable for recycling and made from fully managed and sustained forest sources. Logging, pulping and manufacturing processes are expected to conform to the environmental regulations of the country of origin.

A catalogue record for this book is available from the British Library.

A catalog record for this book is available from the Library of Congress.

*To the
thoughtful,
caring,
diligent,
collaborative,
creative
authors
in my courses,
past and future*

Contents

Acknowledgments   x

About the Author   xi

Contributors   xii

Part I   Creativity Plus Ethics Anticipates A Greater Common Good

1   Origins   3

2   Outlook   15

3   Opportunities   25

Part II   Gadget Controllers

4   Geoengineering: Taking "Change the World" to the Extreme   37
    with DaEun Kim

5   3D Printing: Manufacturing 2.0   45
    with Natalie Spivak

6   Driverless Cars: Driven to Extinction?   53
    with Tomasz Mlodozeniec

7   The Internet of Things: Daily Life, Automated   61
    with Xiaoyi Cui

8   Drones: Insights from Above   69
    with Victoria Westerband

Part III Body Shapers

9	Preventive Gene Testing: Our New Fortune Teller? *with Victoria Russo*	81
10	The Microbiome: Am I a "We"? *with Ria Citrin*	89
11	Stem Cell Therapy: Biological Reboot *with Sarah Schnur*	97
12	Fortified Junk Food: Too Much of a Good Thing? *with Lilia Juarez-Kim*	104
13	Electronic Cigarettes: Huffing and Puffing about Addiction *with Zachary Goodstein*	111

Part IV Emotion Tuners

14	Manipulating Emotions: Feelings for Sale *with Curtis Meyer*	121
15	Legalized Marijuana: The Highs and Lows *with Achu Johnson Alexander*	129
16	Happiness as a Life Goal: The Smiley Face Reigns *with Robert Kagan*	137
17	Boredom Avoidance: Like Watching Paint Dry *with Kaitlin Black*	145

Part V Self Definers

18	Authenticity as a Life Purpose: True to My Self *with Olivia Lourie*	157
19	Gender Fluidity: A Spectrum of Identities *with Sarah Parker*	165
20	Average as Optimum: What Does the "Mean" Really Mean for Me? *with Eliana Hadjiandreou*	174
21	Big Data: What Does It All Add Up To? *with Michael Masters*	182

Part VI Social Connectors

22	The Right to Be Forgotten: Disappearing Acts *with Christopher Charles Canieso*	193
23	Virtual Currency: The Value of Instant Anonymity *with Yonathan Bassal*	200
24	Emoticons: :-) or :-(? *with Jai Sung Lee*	207
25	Digitally Mediated Communication: I Feel So Close to... Who Are You? *with Michaela Hession*	214

Index   223

Acknowledgments

In reverse chronological order: Much thanks to Nicola, Eleanor, Libby, Sharla, and the production team at Palgrave Macmillan for their guidance; the anonymous reviewers for their encouragement and feedback; Kay, Sophie, and Nellie for providing a positive writing environment; Ben for editing and patience; and Tristan for assistance in compiling references. Without the trust, commitment, and imaginations of the student authors, this book would not be. Last but not least, thanks to all the forward-thinking, ethically minded creators who contribute to the positive development of our societies and cultures.

About the Author

Seana Moran, Clark University, USA, received her doctorate in Human Development and Psychology and a Master of Education from Harvard University, as well as a Master of Business Administration, and Bachelor of Arts in Journalism and History. Her research focuses on the intersections of creativity, morality/ethics, life purpose, and wisdom as individuals strive to contribute to their communities. She co-edited *The Ethics of Creativity*, *Multiple Intelligences Around the World* and several volumes of the *Creative Classrooms* series, and co-authored *Creativity and Development*. She has published numerous articles and received several grants, awards, and fellowships. Prior to her scholarly career, she was a practicing "creative" in her own advertising and marketing company.

Contributors

Co-authors were students at Clark University (in alphabetical order):

Achu Johnson Alexander, PhD candidate in developmental psychology; 2008 MA in applied psychology

Yonathan Bassal, 2014 BA with economics/psychology double major

Kaitlin Black, PhD candidate in developmental psychology; 2012 MA in developmental and educational psychology; 2010 BA in psychology and Spanish

Christopher Charles Canieso, 2015 BA with psychology major/management and marketing concentration

Ria Citrin, 2015 BA with psychology/studio art double major

Xiaoyi Cui, 2014 BA with psychology major

Zachary Goodstein, 2015 BA with psychology major/philosophy minor

Eliana Hadjiandreou, 2016 BA with psychology major/cultural studies and communications minor

Michaela Hession, 2015 BA with psychology major

Lilia Juarez-Kim, PhD candidate in developmental psychology; 2014 MA in developmental psychology; 2011 BS in psychology

Robert Kagan, 2014 BA with psychology major/education minor

DaEun Kim, 2015 BA with psychology major/art history minor

Jai Sung Lee, 2015 MS in finance; 2014 BA with psychology major

Olivia Lourie, 2015 BA with communication and culture/studio art double major

Michael Masters, 2014 BA with psychology major/management minor

Curtis Meyer, 2015 BA with psychology/sociology double major

Tomasz Mlodozeniec, 2015 BA with psychology major/communications minor

Sarah Parker, 2015 BA with psychology major/management minor

Victoria Russo, 2016 BA with psychology/sociology double major

Sarah Schnur, 2014 BS with biology/psychology double major

Natalie Spivak, 2015 BS with management/psychology double major

Victoria Westerband, 2015 MBA; 2014 BA with psychology major/management minor

Part I

Creativity Plus Ethics Anticipates A Greater Common Good

1
Origins

Is creativity good or bad?

In the early 21st century, we call on creativity to save lagging economies, solve intractable problems, and offer up entertaining gadgets and lifestyles. We voraciously consume the latest devices, fashion, music, foods, apps, travel destinations, and ideas. We're terrified of not being in the know, behind the times, or perhaps worst of all, obsolete. But creativity was not always held in such high esteem. Where do some of our feelings about creativity originate?

Greek mythology portrays humans who attempt to reorder the way things are as being punished for their hubris. Icarus flew too close to the sun and fell to his death as his wings melted. Prometheus taught humans to use fire and was condemned to birds pecking his liver painfully forever. Sisyphus's clever trickery consigned him to a fate of repeatedly pushing the same boulder uphill. If humans had an idea, it came from divine inspiration, the muses, or our moral spirit – a genie (from which the more positively toned "genius" is derived) or daemon (from which the more sinister "demon" is derived).

Literature depicts creators as parents of abominations, arrogant and destructive loners, and rejected weirdos. In *Frankenstein*, the creation struggles for approval, yet is too different to be accepted into society (Shelley, 1818). In *The Fountainhead*, the creative architect is portrayed as inflexible, willing to destroy his own building rather than have its purity tainted by others (Rand, 1943). In *Jonathan Livingston Seagull*, the bird with his new aerobatic way of flying is shunned by the flock, but faces the scorn by returning to teach the others the power and joy of flight (Bach, 1970). In *The Giver*, alternative ideas are anathema to society as no other options beyond what already has been agreed upon as good are allowed to be thought – only one person in the society, isolated from all others, is the keeper of options from the past, which remain unshared (Lowry, 1993).

History texts recount the triumphs of the few "genius" creators and their creations, such as guns and the atomic bomb in warfare, rule of law and democracy in government, ethics and logic in philosophy, money and markets in economics, perspective and abstraction in arts, genes and the unconscious in sciences, algebra and calculus in mathematics, or the light bulb and airplanes in invention. But primary sources from these geniuses' own eras illuminate creators' struggles, such as Galileo and Copernicus with the Church, or the Impressionists with the Academics in art. Furthermore, individuals and groups whose ideas were not accepted have been left out of this "official" record of the past.

Yet, these descriptions of creativity from the past leave us without practical tools for how we *ought* to think about novel contributions as we encounter them in our lives now. How can we make good judgments about innovations – as creators or users – when they first emerge? The tenets we rely on to judge conventional contributions may not apply to novel contributions. As creativity changes cultures over time, even innovations considered positive at their introduction can lead to long-term negative consequences. For example, understanding bacteria developed antibiotics that save many lives but also led to biological warfare and more virulent bacteria with antibiotic overuse. Or, mortgages allow more people to own homes but also lead to personal bankruptcies and societal financial crises. Conversely, innovations initially criticized as destructive to society – such as slave emancipation or the printing press – eventually were given high regard. Even at the time of introduction, an innovation can differentially affect individuals or groups – for example, those who have a new work-reducing appliance and those who do not, those cured with new treatments and those without access, or entrepreneurs enriched by the computer revolution and workers replaced by robots (Moran, 2012).

Creativity stimulates cultural development

"Creative" has been attributed to an increasing array of individuals, groups, tasks, occupations, processes, environments, ideas, products, and more (see Moran, 2009) to such an extent that it's difficult to discern creative from uncreative. Furthermore, as creativity has grown in economic value, the levels of creativity have expanded into a spectrum from "mini-c" personal understanding through "little-c" everyday creativity to "big-C" eminent creativity featured in history books. I do not dispute the various venues or levels of creativity, which I've discussed elsewhere (Moran, 2015a, 2015b). Instead, in this book, I and the student authors take a developmental view of creativity. Development focuses on the unfurling of potentials into qualitative changes that increase the capacities of a system. In this case, creativity is an endogenous mechanism for cultural development.

Creativity references an individual's or group's contribution to culture that includes a shift of meaning. This meaning-making occurs first by an individual, then is shared with others, some who adopt it and some who reject it. Eventually, the shift may become the new, socially agreed-upon meaning among cultural members (see Bourdieu, 1993; Glaveanu, 2011; Moran, 2009, 2010a, 2010b, 2015b; Moran & John-Steiner, 2003; Valsiner, 2000).

What is at first labeled creative – novel and surprising with potential usefulness (Bruner, 1962) – becomes an innovation when introduced to others. An innovation is a potential tool for renewing some aspect of the culture. If this innovation survives the gauntlet of gatekeepers', critics', and adopters' judgments and is found actually useful (Rogers, 1962; see also Moran, 2015b) within the political, economic, and social affordances of a particular time period (Moran & John-Steiner, 2003; Simonton, 1996), the innovation may become the new cultural norm, worthy to perpetuate to future generations (see Moran, 2014b, 2015b). We may attribute creativity to a person after the fact, as the perceived cause of creativity. But creativity is not something we *are*, it's something we *do*. It is an action that instigates innovation, which develops the culture anew. This view of creativity highlights that creativity is systemic, the system it affects is culture, the effect is by way of meaning development, and the mechanism is via social diffusion.

Creativity is systemic. A system comprises connected parts that interact over time. These interactions lead to dynamics that "move" the system into qualitatively different states. Creativity comprises many systems. Based on Vygotsky's developmental theory (see Moran & John-Steiner, 2003) and Gruber's (1988) evolving systems theory, we adopt agreed-upon meanings from adults or artifacts in our culture. Our purposes, emotions, and knowledge interact to formulate metaphors, insights, and other tools that help us weave learned concepts with imagination into creative thought. Increasingly, our personal experiences develop idiosyncratic *senses* of how the world works. We may share these senses – as personal opinions or insights – with others, who may accept them and pass them on to others (Csikszentmihalyi,

1988, 1999; Moran, 2009, 2015b) through conversation, publication, or symbolization. For example, current social media make these processes more visible and quicker on digital platforms than previous media.

The system affected is culture. Culture comprises the customs tilled and guarded by a group over time. These customs differentiate the group's behavior from other groups' and creates a "bank" of behaviors for future situations that we may encounter (Hofstede, 1980). Customs accumulate through our adoptions from each other of habits, beliefs, and the like (Shweder, 2008). As we interact regularly, we come to share a common set of meaning-making frames. We are all contributors to our culture by continuing valued traditions, often organized through institutions, and by expressing potential improvements to our shared way of living (Moran, 2009).

The currency of culture is meaning. Meaning imbues our experiences with significance, understanding, and a foundation for interpreting later similar experiences (Park, 2010). Culturally agreed-upon meaning instills value that is both personalized and socially sharable – via language and artifacts. Shared meanings can build into collective norms, which make relationships more predictable (Olivares, 2010; Valsiner, 2000). As we develop, the meanings we accept from our culture become habitual. Our contribution is helping to perpetuation the culture by "carrying" cultural knowledge in our minds and behavior.

Of course, we are not mental clones of each other. In any group, variation results from idiosyncratic senses of experiences. This diversity of meaning-making gives rise to uncertainty in social relationships – we aren't really sure what others think (Valsiner, 2000). Most people address this uncertainty by adapting accepted cultural meanings as "good enough" tools for understanding others or by seeking leaders to make sense of ambiguity for them (Mumford et al., 2014).

Creativity contributes new meaning to culture. But sometimes, we may problematize the uncertainty. We consider culture's current ways of addressing a situation as inadequate (Kaufmann, 2004), and we forge our own different sense to restore our personal coherence about our lives (Park, 2010). We may share this idiosyncratic sense with others. Sharing introduces the novel way of thinking into the culture. Our way may alleviate others' uncertainty (Proulx & Heine, 2009). Or it may increase others' uncertainty by showcasing misalignments within the culture (Moran, 2010c) and starting a ripple effect of further sense-making (Moran, 2009).

Creativity turns a cultural tenet into a variable – it can open culture to multiple meanings (Moran, 2014a) as the various ways individuals "make sense of" these meanings compete for attention (Rogers, 1962). Creativity does not require an actual product – simply changing the meaning of an existing object, such as a rock becoming a pet, is sufficient to start a cascade of meaning change. Another example is the notion of "service." It has changed tremendously from personal servants like a lady's maid, to specialized marketed services like hairdressing, to trade services like plumbing, to professional services like medicine, to digital services like social media sites.

Creativity requires acceptance and adoption. Creativity launches innovation, a social judgment process by which cultural members decide the new contribution's worth. An idiosyncratic sense's value is assessed by comparing this new contribution to the culture's norms, capacities, and alternatives at the time (Moran, 2015a). As contributions are accepted, they enrich the repertoire of capabilities and tools by which all of us in the culture can respond to relevant future situations (Vale, Flynn, & Kendal, 2012; Valsiner, 2000). The culture has developed.

At least a few open-minded, tolerant individuals, who are willing to try something before its implications are fully known, are helpful to spread an idiosyncratic sense to even more people (Moran, 2010a). These tolerant "innovators" and "early adopters" help later adopters see the forest for the trees: they shepherd the novel contribution through the trees of individual anxieties to help later adopters visualize the forest of future benefits (Jaques, 1970; Rogers, 1962). Without this early shepherding, the novelty may not spread, and then it cannot develop culture. It becomes an error or a fad that withers away (Moran, 2015b).

Then, the novelty enters the mainstream culture, which is composed of most of the culture's members. Even if we do not instigate cultural development as a creator sharing an unusual personal sense, we still play an important role in renewing culture as adopters. We judge others' contributions when we make purchases, hire employees, vote, join organizations, invest, and learn. These seemingly small acts aggregate to the cultural value of a contribution. As part of this ripple effect, we contribute, in some ways, to a "social activism" by accepting new ideas and meanings that indirectly can alter social structure (Moran, 2010d). For example, the expansion of what constitutes a "person" over time has led to abolition of slavery, women's right to vote, pets viewed as family members, lab chimpanzees argued as liberated beings under the law, and corporations having individual rights.

Creativity involves time and relationship. Judgment and adoption take time as individuals adjust their attention, interest, and investment away from tried-and-true options toward the novel contribution. Surprise at the novelty may stimulate interest, but judgments of usefulness often require recommendations, simulations, or actual use. Immediate technological usefulness is easier to decipher than social and cultural usefulness. Radical innovations are most difficult to judge (Garcia & Calantone, 2002), and often it falls to the creators to persuade others of a novelty's value (Bourdieu, 1993). Adoption rates can be uneven as some of us are more open to innovations, whereas others won't change unless forced to (Rogers, 1962).

Furthermore, adoption can become more difficult when novel introductions come so fast that we don't have the opportunity to "digest" and stabilize past contributions to refer to for judgment criteria (Valsiner, 2000). There is some concern that today's "disruptive innovation" mentality (Christensen, Craig, & Hart, 2001) of nearly constant novel introductions destabilizes our ability to discern creativity from junk. We can't build on each other's contributions because contributions never stabilize to become foundational (Moran, 2015b).

Creativity is expensive for creators or potential adopters. From a cultural-developmental perspective, creativity is far from a safe process, done by a lone genius, with no repercussions. It is not simply play or freedom of self-expression (Moran, 2010b). Creators often have to use their own resources because cultural institutions initially don't know how to support the budding novel contribution nor do other cultural members understand it (Moran, 2015b). Creators may have to go it alone for a while until they find open-minded individuals who can champion their novelty with others (Torrance, 1993). Even then, one creator's contribution may be too different from the current norm and thus it may be rejected as error, or it may be out-maneuvered by another's contribution during the innovation process (Jasper, 2010).

In addition, creativity may waste more cultural resources than tried-and-true methods because of the uncertainty involved. Similarly, adopters may outlay funds or put their reputations on the line for untried offerings that don't work as planned. Throughout history, minds and garbage dumps have been filled with worthless – and even harmful – ideas or products. Perhaps these expenses belie why many people only turn to creativity after more conventional methods fail: it is cheaper to follow the beaten path than to trail-blaze. If what already exists is good enough, why risk valuable resources on possibilities that have much longer odds of being useful?

Creativity and ethics are synergistic

This foray into benefits and harms belies the ethical nature of creativity. Yet, most studies of judgments of creativity do not consider nor include criteria for an ethical dimension. Ethics comprises judgments of good and bad, right and wrong. Is a contribution trustworthy and credible? Does it represent the values of the culture?

Ethics guide us to do the right thing. Although ethics has been studied abstractly as a branch of philosophy, ethics' relation to creativity stems from applied ethics – how people ascertain the rights and wrongs within a particular situation and decide what to do then and there. Ethics not only distills morals into rules for good behavior, it also generates motivations to be good. Ethics codes are useful as baselines, but they can lead to mindless following of the tenets. Blind ethics is dangerous, even if we perfectly conform – especially with creativity and innovation because of their higher uncertainty.

Rather than external codes, it may be more helpful to characterize ethics as an internal compass that steers us toward "the good" and keeps us from diverting our attention based on immediate incentives or distractions (Cua, 1978; Weston, 2013). Thus, our ethics – how we affect others – are part of our purpose directing our lives (Moran, 2014c). The particular criterion we use for justifying our or others' behavior – such as the ends justifying the means, or what's best for the most people, or what our duties to others entail (Weston, 2013) – may be less important than our sensitivity to the specific ethical implications of situations we encounter (Narvaez & Endicott, 2009).

Accountability blames and punishes. One framing is that ethics determines blame and restitution after an infraction occurs: it is accountability. But this framing focuses on what we do wrong, not what we do right. This focus may reinforce a negative view of humanity – that we should not trust each other – which can spiral into social isolation or violent chaos because we forget how to cooperate. Furthermore, this framing is retrospective and does not allow ethics a proactive role in the development of character or society. Once we've "paid our dues" for an infraction, there is no more guidance to move forward. Even understanding the underlying values on which ethical tenets are built – such as self-interest, relationships, or principles (Kohlberg, 1981); care, reciprocity, loyalty, respect, or purity (Haidt & Graham, 2007); or desire, obligation, and the sacred (Shweder, 2008) – does not provide us clarity on how to behave in a future situation.

Responsibility personalizes ethics. Instead, what is called for is a way to transform the ethos – the culture's "spirit of the times" including its values – into our own personal lens. Just as cultural meaning in knowledge domains or experience is adopted and adapted in a partially idiosyncratic way by "making sense of" the meaning, cultural ethics can be adopted and adapted through "making us responsible for" the situation. Whereas accountability focuses on settling accounts based on past acts, responsibility emphasizes responding to the situation appropriately – "response-ability." Responsibility is situated, contextualized, proactive, and future-oriented – what can I do here that is good now for those involved, and sets a good direction for others in the future as well (Weston, 2013)?

Rather than singling out ourselves as the center of our lives, we recognize we are part of a tapestry of individuals and institutions that contribute to our collective momentum. In this case, momentum might be described as the multiplicative effect of individuals' various paces of life and the significance they ascribe to life. From this view, ethics is a hopeful, interactive way for the variety of valid perspectives contributing to our joint efforts to benefit from our mutual contributions. Ethics provides a "threading" of our individual life strands in the tapestry, and diversity of perspectives generates more colorful patterns and possibilities. If we are more aware, insightful, and anticipatory about the contributions we make, ethics is not a punishment nor a chore. It is an opportunity.

Creativity and ethics in the abstract offer little help during "moments of truth." It has been difficult for researchers to determine how creativity and ethics, as abstractions, relate to each other. So far, ethicists and moral developmentalists suppose stable cultures and situations as backgrounds for a particular ethical dilemma. They ask people to ponder only the options presented in the dilemma, much like multiple-choice questions, with no room for other possible responses (Moran, 2014b). Similarly, creativity and innovation research often suppose creativity to be amoral (mostly through simply not mentioning the ethical dimension at all). Creativity is beyond judgments of good and bad because no clear ethical criteria exist for novel contributions (Bourdieu, 1993). When ethics and creativity are brought together, they are posited as opposites: ethics represents duties and rules, and creativity represents freedom and breaking rules.

Although several psychology papers address a moral dimension to creativity in the abstract (Gruber, 1993; Runco, 1993; Runco & Nemiro, 2003), the goodness or badness of creativity on more concrete terms only recently has become a topic of interest (Cropley et al., 2010; Gino & Ariely, 2012; Moran, Cropley, & Kaufman, 2014). These approaches still suffer from decontextualization in lab experiments or retrospection in case studies. Furthermore, many of them assume that creativity and ethics are antagonistic and, thus, creativity is *un*ethical. Thus, the relationship between creativity and ethics is still assumed, rather than explored.

"Ought implies can implies create"

This quote was the challenge given to us more than 20 years ago (Gruber, 1993). Ethics impels creativity, and creativity demands ethics. They compose each other, like M. C. Escher's famous "Drawing Hands." The two are symbiotic, their existence interwoven for mutual benefit. Similarly, opportunities are the flip side of challenges. Our opportunity, in this book, is exploring this: What will help us in the real world to intersect creativity's "what if?" with ethics' "why if?" How do we move forward?

Beyond novelty, usefulness, and surprise, will we find the good? In addition to the US Patent Office's criteria (Simonton, 2012), should creativity be judged on a criterion of goodness? This ethical dimension recognizes that a novel contribution impacts lives and societies by introducing benefits and harms to others or to the collective, and these effects may unfold over time. Furthermore, creativity may generate duties for creators and adopters to consider how the novel contribution might be used or misused and how the novel contribution may spawn interpersonal, social, and cultural effects beyond its initial purpose (Moran, 2010c, 2014a, 2014b). For example, eating utensils changed the shape of the human mouth and led to table manners; the theory of relativity influenced narrative structure; and the automobile restructured our landscapes and social engagements (Tenner, 1996). Creativity and innovation ripple through culture as a force of change, potentially restructuring cultural dynamics (Moran, 2009, 2014b). Is that good?

References

Bach, R. (1970). *Jonathan Livingston Seagull*. New York, NY: Macmillan.
Bourdieu, P. (1993). *The field of cultural production*. New York, NY: Columbia University Press.
Bruner, J. S. (1962). The conditions of creativity. In J. S. Bruner, *On knowing: Essays for the left hand* (pp. 17–30). New York, NY: Cambridge University Press.
Christensen, C., Craig, T., & Hart, S. (2001, March/April). The great disruption. *Foreign Affairs*, 80–95.

Cropley, D. H., Cropley, A. J., Kaufman, J. K., & Runco, M. A. (2010). *The dark side of creativity*. New York, NY: Cambridge University Press.
Csikszentmihalyi, M. (1988). Society, culture, and person: A systems view of creativity. In R. J. Sternberg (Ed.), *The nature of creativity* (pp. 325–339). New York, NY: Cambridge University Press.
Csikszentmihalyi, M. (1999). Implications of a systems perspective for the study of creativity. In R. J. Sternberg (Ed.), *Handbook of creativity* (pp. 313–338). Cambridge, UK: Cambridge University Press.
Cua, A. S. (1978). *Dimensions of moral creativity: Paradigms, principles, and ideals*. University Park, PA: The Pennsylvania State University Press.
Garcia, R., & Calantone, R. (2002). A critical look at technological innovation typology and innovativeness terminology: A literature review. *Journal of Product Innovation Management, 19*(2), 110–132.
Gino, F., & Ariely, D. (2012). The dark side of creativity: Original thinkers can be more dishonest. *Journal of Personality and Social Psychology, 102*(3), 445–459.
Glaveanu, V. P. (2011). Creativity as cultural participation. *Journal for the Theory of Social Behaviour, 41*(1), 48–67.
Gruber, H. E. (1988). The evolving systems approach to work. *Creativity Research Journal, 1*(1), 27–51.
Gruber, H. E. (1993). Creativity in the moral domain: Ought implies can implies create. *Creativity Research Journal, 6*(1–2), 3–15.
Haidt, J., & Graham, J. (2007). When morality opposes justice: Conservatives have moral intuitions that liberals may not recognize. *Social Justice Research, 20*, 98–116.
Hofstede, G. (1980). Motivation, leadership and organization: Do American theories apply abroad? *Organizational Dynamics, 9*(1), 42–63.
Jaques, E. (1970). *Work, creativity, and social justice*. New York, NY: International University Press.
Jasper, J. M. (2010). The innovation dilemma: Some risks of creativity in strategic agency. In D. H. Cropley, A. J. Cropley, J. C. Kaufman, & M. A. Runco (Eds.), *The dark side of creativity* (pp. 91–113). New York, NY: Cambridge University Press.
Kaufmann, G. (2004). Two kinds of creativity – but which ones? *Creativity and Innovation Management, 13*(3), 154–165.
Kohlberg, L. (1981). *Essays on moral development, Vol. I: The philosophy of moral development*. New York, NY: Harper & Row.
Lowry, L. (1993). *The giver*. Boston, MA: Houghton Mifflin.
Moran, S. (2009). Creativity: A systems perspective. In T. Richards, M. Runco, & S. Moger (Eds.), *The Routledge companion to creativity* (pp. 292–301). London, UK: Routledge.
Moran, S. (2010a). Changing the world: Tolerance and creativity aspirations among American youth. *High Ability Studies, 21*(2), 117–132.
Moran, S. (2010b). Creativity in school. In K. S. Littleton, C. Wood, & J. K. Staarman (Eds.), *International handbook of psychology in education* (pp. 319–360). Bingley, UK: Emerald Group.
Moran, S. (2010c). Returning to the Good Work Project's roots: Can creative work be humane? In H. Gardner (Ed.), *Good Work: Theory and practice* (pp. 127–145). Cambridge, MA: Good Work Project.

Moran, S. (2010d). The roles of creativity in society. In J. C. Kaufman & R. J. Sternberg (Eds.), *The Cambridge handbook of creativity* (pp. 74–90). New York, NY: Cambridge University Press.
Moran, S. (2012). Book review: The dark side of creativity (D. H. Cropley, A. J. Cropley, J. C. Kaufman & M. A. Runco, eds.). *Psychology of Aesthetics, Creativity, and the Arts, 6*(3), 295–296.
Moran, S. (2014a). An ethics of possibility. In S. Moran, D. H. Cropley, & J. C. Kaufman (Eds.), *The ethics of creativity* (pp. 281–298). Basingstoke, UK: Palgrave Macmillan.
Moran, S. (2014b). The crossroads of creativity and ethics. In S. Moran, D. H. Cropley, & J. C. Kaufman (Eds.), *The ethics of creativity* (pp. 1–22). Basingstoke, UK: Palgrave Macmillan.
Moran, S. (2014c). What "purpose" means to youth: Are there cultures of purpose? *Applied Developmental Science, 18*(3), 1–13.
Moran, S. (2015a). Adolescent aspirations for change: Creativity as a life purpose. *Asia Pacific Education Review, 16*(2), 167–175. doi: 10.1007/s12564-015-9363-z
Moran, S. (2015b). Creativity is a label for the aggregated, time-dependent, subjective judgments by creators and adopters. *Creativity: Theories-Research-Applications, 2*(1). Available at: http://www.creativity.uwb.edu.pl/index.php/en/creativity-2-1-2015.
Moran, S., & John-Steiner, V. (2003). Creativity in the making: Vygotsky's contribution to the dialectic of creativity and development. In K. Sawyer et al., *Creativity and development* (pp. 61–90). New York, NY: Oxford University Press.
Moran, S., Cropley, D. H., & Kaufman, J. C. (2014). *The ethics of creativity*. Basingstoke, UK: Palgrave Macmillan.
Mumford, M. D., Peterson, D. R., MacDougall, A. E., Zeni, T. A., & Moran, S. (2014). The ethical demands made on leaders of creative efforts. In S. Moran, D. H. Cropley, & J. C. Kaufman (Eds.), *The ethics of creativity* (pp. 265–278). Basingstoke, UK: Palgrave Macmillan.
Narvaez, D., & Endicott, L. G. (2009). *Ethical sensitivity: Nurturing character in the classroom*. Notre Dame, IN: ACE Press.
Olivares, O. J. (2010). Meaning making, uncertainty reduction, and the functions of autobiographical memory: A relational framework. *Review of General Psychology, 14*(3), 204–211.
Park, C. L. (2010). Making sense of the meaning literature: An integrative review of meaning making and its effects on adjustment to stressful life events. *Psychological Bulletin, 136*(2), 257–301.
Proulx, T., & Heine, S. J. (2009). Connection from Kafka: Exposure to meaning threats improves implicit learning of artificial grammar. *Psychological Science, 20*(9), 1125–1131.
Rand, A. (1943). *The fountainhead*. Indianapolis, IN: Bobbs-Merrill.
Rogers, E. M. (1962/1983). *Diffusion of innovations, 3rd ed.* New York, NY: Free Press.
Runco, M. A. (1993). Creative morality: Intentional and unconventional. *Creativity Research Journal, 6*(1–2), 17–28.
Runco, M. A., & Nemiro, J. (2003). Creativity in the moral domain: Integration and implications. *Creativity Research Journal, 15*(1), 91–105.
Shelley, M. (1818). *Frankenstein*. London, UK: Lackington, Hughes, Harding, Mavor & Jones.

Shweder, R. A. (2008). Why cultural psychology? *Ethos, 27*(1), 62–73.
Simonton, D. K. (1996). Individual genius within cultural configurations: The case of Japanese civilization. *Journal of Cross-Cultural Psychology, 27*(3), 354–375.
Simonton, D. K. (2012). Taking the U.S. Patent Office criteria seriously. *Creativity Research Journal, 24*(2–3), 97–106.
Tenner, E. (1996). *Why things bite back: Technology and the revenge of unintended consequences.* New York, NY: Vintage.
Torrance, E. P. (1993). The beyonders in a thirty year longitudinal study of creative achievement. *Roeper Review, 15*, 131–134.
Vale, G. L., Flynn, E. G., & Kendal, R. L. (2012). Cumulative culture and future thinking: Is mental time travel a prerequisite to cumulative cultural evolution? *Learning and Motivation, 43*, 220–230.
Valsiner, J. (2000). *Culture and human development.* Thousand Oaks, CA: Sage.
Weston, A. (2013). *A 21st century ethical toolbox, 3rd ed.* New York, NY: Oxford University Press.

2
Outlook

Our outlook conveys where we direct our focus as we navigate into the future as adopters, and perhaps invent the future as creators. If what we ought to do is to create, how might we apply the connections made between creativity and ethics in the Origins chapter? This chapter explores our ethical worries and hopes regarding creativity. Worry dreads a negative outcome, whereas hope strives for a better outcome.

Focusing on risk

We are worriers. We tend to spend more energy avoiding potential losses than pursuing potential gains. "Loss aversion" is a well-documented bias in human thinking (Tom, Fox, Trepel, & Poldrack, 2007). We dislike reminders in our everyday life of the risks inherent in change (Stacey, 1996). We worry that novel contributions may not be better than what already exists, or that their costs may outweigh their benefits, or that they may be difficult to learn how to use.

Even if innovations work as expected, their novelty can create social turbulence as individuals adopt them at different rates. While the inconsistencies and difficulties of innovation are worked out, misalignments can occur between expectations and reality such as between professionals' standards and their actual performance. For example, patients might see more diversity in cure rates during the period when new treatments are launched, as doctors learn the new protocols (Moran, 2010c).

Whereas prior to an innovation's introduction it may have been easier to discern truth from lie, the new meaning added to culture by the innovation makes assessments of honesty more difficult. Creators or adopters of the innovation have widened their repertoire of meaning, perhaps even beyond what some subcultural groups sanction. Terminology shifts and confusion can increase as embracers of novelty try to communicate with adherents to tradition. As a result, the avant-garde meaning-makers may be considered less honest by the cultural members who still abide by meanings from before the innovation (Gino & Ariely, 2012; Kunzendorf & Bradbury, 1983; Moran, 2014a). For example, since social media was introduced, meanings for terms such as "friend" and "tweet" are much different for teens than for the elderly.

Even if creators seem trustworthy, we adopters may wonder if they have our interests in heart. We may not understand how innovations work – for example, the "behind the scenes" software code of social media sites – so we have few mechanisms to assess creators' motives. Some research suggests that professionals sometimes do not consider longer-term uses or impacts of their work on people whom they've not met (Gardner, Csikszentmihalyi & Damon, 2001). For example, geneticists may not think about the unborn third generations of families with hereditary diseases, and newspaper journalists may not think about the people in small villages who cannot read. Furthermore, young professionals fret about their own success as a precursor to being ethical: that is, many believe that they must first "make it" in their chosen field and, after that, they will be in a position to be ethical (Fischman, Solomon, Greenspan, & Gardner, 2004).

Especially for people who resist adopting an innovation, the innovation's existence can feel like an imposition or aggression. The innovation can feel *wrong*. It can upset formerly stable relationships and cultural standards, which can spill over into situations completely unrelated to the innovation's original purpose (see Moran, 2014b). The widening gap between the creators and early adopters versus the laggards who refuse to adopt an innovation (Rogers, 1962) may trigger disgust in each other as the values each holds dear diverge from the other group's values (Haidt & Graham, 2007; Rozin, Lowery, Imada, & Haidt, 1999). Perhaps worst of all, we worry that the "aberrant" behaviors the innovation causes may spread, and those who do not adopt the new way become the minority who didn't keep up with progress.

Many studies show, in general, leaders tend to have a bias *against* creativity. Although we say we want more creativity in the world to improve our way of life, many of us do not actually want to experience the effects creativity causes in society. We don't like being experimented on, or inconveniences in services, or wastefulness – all which can come with creativity's inefficient trial-and-error process. Furthermore, most people feel more comfortable around like-minded people rather than original thinkers (see Moran, 2010a). It is as if we can't allow ourselves to stop worrying about what may happen next. Creativity is associated with uncertainty, uncertainty breeds anxiety, anxiety provokes vigilance (Jaques, 1955). Perhaps to try to avoid this spiral of worry, some of us simply make an a priori judgment that creativity is not good. It is something to avoid if we can.

Perceiving brighter horizons

Worrying about what could go wrong has merits. It helps keep us safe from harm. But it also can stall us from pursuing good opportunities, which is why leaders of creative endeavors often focus on "sensemaking" of novelties and opportunities (Mumford, Peterson, MacDougall, Zeni, & Moran, 2014). Worrying traps us in a cramped world of the known, where we may feel comfortable in part because we refuse to look toward the horizons of hope. Hope is not optimism. Optimists think the situation is fine, regardless of what the situation actually is. They refuse to concede if something is sub-par (Izuma & Adolphs, 2011). Thus, optimism may not motivate us to make the situation better. We accept it as it is, and we convince ourselves it's fine. Hope, on the other hand, recognizes that there is a gap between the current situation and the way we'd like it to be. Like a stretched rubber band, there is some tension in hope between what is and what could be, and that tension motivates us to work toward our ideal.

Hope is driven by moral imagination. Moral imagination allows us to envision new opportunities to serve others or the common good, to respect and empathize with others' perspectives, and to rehearse potential options before acting on them in the real world (Fesmire, 2003; Johnson, 1993; Narvaez & Mrkva, 2014). Furthermore, moral imagination allows the future to play a role in our *present* decision-making: our vision of tomorrow can give direction to the way we act today (Seligman, Railton, Baumeister, & Sripada, 2013).

Moral imagination provides symbolic tools to consider potential repercussions of our choices in advance, thus keeping ourselves and others safe from our potential errors, while at the same time also expanding the scope of the actions we might engage. We can see the world through others' eyes and choose from a wider repertoire of responses. With imagination as our moral guide, we are not limited to what others have passed down to us. Therefore, we might be more flexible in addressing ethical issues in a specific situation.

Furthermore, we can devise metaphors to scaffold understandings of emerging situations and revise moral narratives with more desirable outcomes (Haste & Abrahams, 2008). Instead of cobbling together stop-gap fixes to moral quandaries based on externally imposed ethical rules, moral imagination pioneers proactive designs for a more coherent ethical practice (Cua, 1978; Weston, 2007).

Caring for the common good

Often, we consider ethics as caring for other individuals. But it also involves taking care of the community, social structure, and values that connect individuals. Ethics is a guardian of the "common good." The common good comprises resources shared and cherished by all cultural members, but which can be depleted through overuse or abuse of those resources (Etzioni, 2004).

Classic analyses of the common good posit the loss of these scarce resources as a tragedy (Hardin, 1968). However, perhaps the tragedy is the assumption that the common good is simply a resource to consume (Moran, 2014b). The common good is not a storage facility. Rather, the common good is like a network that keeps what is good and valued in circulation among cultural members. It comprises situations and behaviors that allow us to perceive, mimic, and perpetuate ethical possibilities. The more we engage each other ethically, the more ethical we all become. We lose the moral myopia born of solipsistic self-interest and expand our ethical horizon to incorporate more and more "others."

Prosociality correlates with originality (Grant & Berry, 2011), and creators seem particularly adept at widening horizons of what is considered good (Havel, 1997; Moran, 2009). Thus, creativity may be a particularly virile tool for developing the common good. Here I outline a few ways this creative development of the common good might occur (Moran, 2010b).

Creativity gives "unspeakable" issues access to the common good. Creativity can "open up" the common good to wider opportunities to care for each other by giving voice to previously taboo topics. Creative fields – especially the arts – give license to set aside our everyday social roles or identities and avail ourselves to other possibilities. Without the "allowance" that a play space offers – such as through music, improvisation, jokes, or costumed festivals – we might not otherwise be able to express some ideas initially. Once the silence has been breached, other voices and perspectives can join the dialogue. For example, graphic artist Shepard Fairey's OBEY stickers and street art call attention to mindless acceptance of propaganda. In the 1980s and 1990s, a group of anonymous artists became the Guerilla Girls, gorilla-masked avengers of racism and sexism via humor and posters. Through the playfulness, these creative approaches can introduce serious issues to the common good that may be too difficult or dangerous to convey in undisguised form because of imbalances of power (Scott, 1990).

Creativity involves wider arrays of individuals and venues in the common good. Unusual portrayals of ethical issues, such as through poetry or theater or mathematical equations, educate cultural members to alternative conceptions of what is good. For example, the Theater of the Oppressed uses audience members as "spect-actors" to explore, analyze, and transform ethically fraught situations as they unfold (Boal, 1993). These alternative venues help people, who previously did not see themselves as cultural producers, characterize themselves as agents of influence. They can experience the fruits of their efforts, which provide a powerful feedback loop for ethical behavior.

Creativity presents images for what the common good itself can become. An extreme position is that creativity is the purpose of the common good. The purpose is devising a social process that can perpetuate further transformation and support institutional flexibility to become more responsive to changing circumstances. Historical examples include the printing press's transformation of how people could relate to the Word of God, the American Constitution's tenets to adapt to and instigate forms of political relationship, and Gandhi's nonviolent civil disobedience methods to change group relations. More recent examples are often technological, such as the Internet's framework to revise the relationships between people and information. Creativity in care of the common good epitomizes what developmental psychologist David Feldman calls our "transformational imperative" to push beyond current bounds, or perhaps more accurately, to evolve beyond our current circumstance (Ambrose, 2014).

Ethical anticipation

Creativity is dependent on time: it has duration. Novel contributions do not appear instantaneously. The roles of foresight and forecasting have been studied in relation to creativity, although research demonstrates that these visions of the future are usually only vague outlines, not fully formed conceptions. We don't get it right in the details (Simonton, 2012). But just because foresight is not 20/20 does not make it irrelevant. Even moderate consideration of how today's actions make or break opportunities tomorrow can correct our moral myopia (Moran, 2014a).

We may not be able to accurately *predict* what will happen tomorrow. Besides, prediction sounds technical, statistical, lacking in the emotional engagement that ethics calls for. A future-oriented ethics calls for anticipation, which not only states an expectation of what could be, but also asserts hope and promise in that expectation for the future. Anticipation doesn't just acknowledge the long view, it celebrates it.

Anticipation of emotions may be particularly powerful for strengthening our ethical sensitivity. Emotions stimulate us, in later relevant situations, to remember past harms or benefits (Baumeister, Vohs, DeWall, & Zhang, 2007). We may not recall the specifics of the past ethics-relevant situation, but the emotional resonance still lets us know to be aware of our effects on others. We *feel* the right thing to do long before we *know* why because the emotion triggers us to think more carefully.

For example, savoring a future reward can boost the enjoyment we derive by stretching the pleasure over the period between now and consummation of the pleasure, as we build up our anticipation of the actual event (Loewenstein, 1987). On the other hand, if we dread what is coming, we can choose to get the event or task over with sooner, thus shortening our suffering. Or expected regret if we pursue a particular course of action gives us a painful signal *now*, when we can do something to avoid the cost that the course of action may incur (Sarangee, Schmidt, & Wallman, 2013).

One reason anticipation may be fruitful for ethics is because ethics too often is not considered until the end of a task, when implementation is underway. This is also true for tasks involving creativity (Mumford, Waples, Antes, Brown, et al., 2010). The implementation stage may be too late because creators already have committed to – and may be determined to succeed on – their current course. Earlier stages in the creative process are easier to induce deliberation about which direction creators should go, thus short-circuiting poorly considered determination to proceed on a course that may lead to worse outcomes (Taylor & Gollwitzer, 1995).

Moral imagination contemplates broader horizons. Ethical anticipation plans for better outcomes. Together, they provide insights beyond the here-and-now to consider consequences addressing wider wheres-and-whens, for whoms, and hows. Creative ethics and ethical creativity decenter us from our own isolated well-being. We recognize ourselves not only as independent beings but also as sustainers of a common good. Creativity and innovation invent the future. The future is the direction we are all headed, so we might as well be more ardent in our progress toward making this collective future as bright as possible.

Standing the test of time

For a novel contribution to survive in a culture requires it to be useful and good for the culture's members. Although we tend to characterize the most creative contributions as what is "far out" or "radical," the truly *most* creative are the contributions that later in time become taken for granted as so *ordinary*. The most creative contributions to culture are the ones that endure.

For example, the number zero. Think about it: zero is a way to represent nothingness or emptiness. Invented in India, it made its way to Europe by way of Arab travelers and the Moors' invasion of Spain (Kaplan, 2000). Zero is fundamental to keeping accurate accounts, calculating outcomes of equations, graphing, and determining concepts of motion – such as speed or acceleration in any given instance. Without zero, engineering, finance, economics, computer programming – and all the other industries that rely on these fields – would be nearly impossible.

Another example: the concept and measure of time. Early humans observed movement and change in the days, seasons, and other rhythms of life. With agriculture, we started to plan our movements to coincide with some of these rhythms. But two tasks occurring together require coordination in time. So we developed clocks – first primitive sundials, then mechanical clocks with weights, springs, or pendulums that measured regular periods, then quartz, digital, and atomic clocks that improved accuracy in the correspondence between measured time and the earth's actual rotation. The railroads standardized time by putting everyone in the US on a singular clock – which was not without controversy as, before then, each town had its own official time. Then in the early 20th century, the railroads convinced the government to institute standardized time zones to facilitate interstate commerce (Landes, 2000). The ethics of daylight savings time is still debated today, and not all states observe it.

A final example: eating utensils. Chopsticks were created thousands of years ago in China to help cooks take food out of cooking pots. Around 400 BC, cooks conserved fuel by chopping ingredients into small pieces that would cook faster, so knives were not needed. Chopsticks spread to Japan, Korea, and Vietnam early in the common era (Bramen, 2009b). Forks, on the other hand, developed in Egypt from the trouble that eaters had holding meat with only a knife. At first, they were considered sinful because the Church considered hands the natural eating instruments. But fork use spread to Europe and ushered in table manners as well as table fashions (Bramen, 2009a).

Thus, when considering whether or how creative and how ethical a novel introduction is, it would be interesting to use time-tested innovations – such as zero, time, and forks – as benchmarks. Novel contributions usually are responses to an existing problem. But these contributions, in turn, create new problems. Over history, we have designed ourselves into and out of difficulty time and again (Thackara, 2005). Intersecting the concepts and examples of creativity and ethics continues this cycle.

When contemplating the ethical implications of creative endeavors – whether through the cases in this book or in our own lives and communities – it is important for us to think repeatedly about the human complexities that can affect how a novel contribution may be accepted, rejected, repurposed, interact with other ideas or products introduced before and after its own introduction, and perhaps persevere to stay in use for centuries. Rarely is this path linear or smooth. More often, it involves loops, switchbacks, gaps, and other obstacles. This book's cases give you the opportunity to practice such thinking by applying creative skills, moral imagination, and ethical anticipation to innovations that emerged into the US mainstream culture in the 2010s.

References

Ambrose, D. (2014). Transformational imperative for the field: An interview with David Henry Feldman. *Roeper Review, 36*(4), 207–209.

Baumeister, R. F., Vohs, K. D., DeWall, C. N., & Zhang, L. (2007). How emotion shapes behavior: Feedback, anticipation, and reflection, rather than direct causation. *Personality and Social Psychology Review, 11*(2), 167–203.

Boal, A. (1993). *Theater of the oppressed*. New York, NY: Theatre Communications Group.

Bramen, L. (2009a, July 31). A history of western eating utensils, from the scandalous fork to the incredible spork. Smithsonian.com

Bramen, L. (2009b, August 5). The history of chopsticks. Smithsonian.com

Cua, A. S. (1978). *Dimensions of moral creativity: Paradigms, principles, and ideals*. University Park, PA: The Pennsylvania State University Press.

Etzioni, A. (2004). *The common good*. Cambridge, UK: Polity.

Fesmire, S. (2003). *John Dewey and moral imagination*. Bloomington, IN: Indiana University Press.

Fischman, W., Solomon, B., Greenspan, D., & Gardner, H. (2004). *Making good: How young people cope with moral dilemmas at work*. Cambridge, MA: Harvard University Press.

Gardner, H., Csikszentmihalyi, M., & Damon, W. (2001). *Good work: When excellence and ethics meet*. New York, NY: Basic Books.

Gino, F., & Ariely, D. (2012). The dark side of creativity: Original thinkers can be more dishonest. *Journal of Personality and Social Psychology, 102*(3), 445–459.

Grant, A. M., & Berry, J. W. (2011). The necessity of others is the mother of invention: Intrinsic and prosocial motivations, perspective taking, and creativity. *Academy of Management Journal, 54*(1), 73–96.

Haidt, J., & Graham, J. (2007). When morality opposes justice: Conservatives have moral intuitions that liberals may not recognize. *Social Justice Research, 20*, 98–116.

Hardin, G. (1968). The tragedy of the commons. *Science, 162*, 1243–1248.

Haste, H., & Abrahams, S. (2008). Morality, culture and the dialogic self: Taking cultural pluralism seriously. *Journal of Moral Education, 37*(3), 377–394.

Havel, V. (1997). *The art of the impossible: Politics as morality in practice*. New York, NY: Knopf.

Izusa, K., & Adolphs, R. (2011). The brain's rose-colored glasses. *Nature Neuroscience, 14*(11), 1355–1356.

Jaques, E. (1955). Social systems as a defense against persecutory and depressive anxiety. In M. Klein, P. Heimann, & R. E. Money-Kyrle (Eds.), *New directions in psychoanalysis* (pp. 478–498). London, UK: Tavistock.

Johnson, M. (1993). *Moral imagination: Implications of cognitive science for ethics*. Chicago: University of Chicago Press.

Kaplan, R. (2000). *The nothing that is: A natural history of zero*. New York, NY: Oxford University Press.

Kunzendorf, R. G., & Bradbury, J. L. (1983). Better liars have better imaginations. *Psychological Reports, 52*, 634.

Landes, D. (2000). *Revolution in time: Clocks and the making of the modern world*. Cambridge, MA: Belknap Press.

Loewenstein, G. (1987). Anticipation and the valuation of delayed consumption. *The Economic Journal, 97*(387), 666–684.

Moran, S. (2009). What role does commitment play among writers with different levels of creativity? *Creativity Research Journal, 21*(2–3), 243–257.

Moran, S. (2010a). Creativity in school. In K. S. Littleton, C. Wood, & J. K. Staarman (Eds.), *International handbook of psychology in education* (pp. 319–360). Bingley, UK: Emerald Group.

Moran, S. (2010b, August). Creativity in the commons. Presentation at the Second Institute on Civic Studies, Tufts University, Medford, MA.

Moran, S. (2010c). Returning to the Good Work Project's roots: Can creative work be humane? In H. Gardner (Ed.), *Good work: Theory and practice* (pp. 127–145). Cambridge, MA: Good Work Project.

Moran, S. (2014a). An ethics of possibility. In S. Moran, D. H. Cropley, & J. C. Kaufman (Eds.), *The ethics of creativity* (pp. 281–298). Basingstoke, UK: Palgrave Macmillan.

Moran, S. (2014b). The crossroads of creativity and ethics. In S. Moran, D. H. Cropley, & J. C. Kaufman (Eds.), *The ethics of creativity* (pp. 1–22). Basingstoke, UK: Palgrave Macmillan.

Mumford, M. D., Peterson, D. R., MacDougall, A. E., Zeni, T. A., & Moran, S. (2014). The ethical demands made on leaders of creative efforts. In S. Moran, D. H. Cropley, & J. C. Kaufman (Eds.), *The ethics of creativity* (pp. 265–278). Basingstoke, UK: Palgrave Macmillan.

Mumford, M. D., Waples, E. P., Antes, A. L., Brown, R. P., Connelly, S., Murphy, S. T., & Devenport, L. D. (2010). Creativity and ethics: The relationship of creative and ethical problem-solving. *Creativity Research Journal, 22*(1), 74–89.

Narvaez, D., & Mrkva, K. (2014). The development of moral imagination. In S. Moran, D. H. Cropley, & J. C. Kaufman (Eds.), *The ethics of creativity* (pp. 25–46). Basingstoke, UK: Palgrave Macmillan.

Rogers, E. M. (1962/1983). *Diffusion of innovations, 3rd ed.* New York, NY: Free Press.

Rozin, P., Lowery, L., Imada, S., & Haidt, J. (1999). The CAD triad hypothesis: A mapping between three moral emotions (contempt, anger, disgust) and three moral codes (community, autonomy, divinity). *Journal of Personality and Social Psychology, 76*, 574–586.

Sarangee, K., Schmidt, J. B., & Wallman, J. P. (2013). Clinging to slim chances: The dynamics of anticipating regret when developing new products. *Journal of Product Innovation Management, 30*(5), 980–993.

Scott, J.C. (1990). *Domination and the arts of resistance: Hidden transcripts.* New Haven, CT: Yale University Press.

Seligman, M. E. P., Railton, P., Baumeister, R. F., & Sripada, C. (2013). Navigating into the future or driven by the past. *Perspectives on Psychological Science, 8*(2), 119–141.

Simonton, D. K. (2012). Foresight, insight, oversight, and hindsight in scientific discovery: How sighted were Galileo's telescopic sightings? *Psychology of Aesthetics, Creativity, and the Arts, 6*(3), 243–254.

Stacey, R. D. (1996). *Complexity and creativity in organizations.* San Francisco, CA: Berrett-Koehler.

Taylor, S. E., & Gollwitzer, P. M. (1995). Effects of mindset on positive illusions. *Journal of Personal and Social Psychology, 69*(2), 213–226.

Thackara, J. (2006). *In the bubble: Designing in a complex world.* Cambridge, MA: MIT Press.

Tom, S. M., Fox, C. R., Trepel, C., & Poldrack, R. A. (2007). The neural basis of loss aversion in decision-making under risk. *Science, 315*(5811), 515–518.

Weston, A. (2007). *Creative problem-solving in ethics.* New York, NY: Oxford University Press.

3
Opportunities

Now it's your turn!

An opportunity is a favorable circumstance for something to happen. This book's aim is to provide a favorable circumstance for you to practice creativity, ethics, and responsibility – not only practice in the sense of pursuing mastery of these skills, but also practice in the sense of setting good habits for yourself. Having made your own sense of the ideas in the prior two chapters, it's time for you, first, to step into the role of an "ethicist of possibilities" who anticipates ways to direct novel contributions toward making the most for the common good, and second, to assume your rightful place as a responsible contributor to our culture.

The following 22 cases feature innovations that have recently emerged into the mainstream culture. These innovations may have been accessible to experts, professionals, or other special groups for a while. They may have been in development over years in special "skunk works," laboratories, or other organizations. Technical reports or specialist blogs can be found about them. But only in the last few years have the ideas, inventions, or procedures launched into major media and "gone viral" in everyday conversations.

This timing of recent emergence into the mainstream is important. It is a critical point in a contribution's lifespan when creativity transforms into innovation. Think of cultural development like a roller coaster. A creator's idiosyncratic sense of some aspect of culture spawned creativity, and that personal understanding is shared. This is like the novel contribution climbing into a car (a cultural field) on the roller coaster. The car moves slowly along a flat or slightly uphill incline for a little while, as it spreads to others in the same roller coaster car (that is, in the same field).

Then, seemingly all of a sudden, the car screeches, lurches, and starts to chug upward. The novel contribution is going to new heights where it will be even more widely visible – it is gaining traction and momentum on the "social circuit" that can make more people aware of the novel contribution. It has hit the "big time," the mainstream culture. Near that moment of lurching upward is the moment we call a contribution an "emerging innovation" – when talk about the contribution is accelerating, but not yet to the point when it seems like *everyone* is familiar with it.

Books take time to write and produce. In the two years this book was developed (2013–2015), four innovations took off. They became mainstream phenomena, usually as the result of laws or regulations being passed or implemented: virtual currency and legalized marijuana in winter/spring 2014, emoticons/emojis and gender fluidity in spring 2015. As this book went to press, driverless cars were poised to take off as they were allowed onto actual roads with human drivers. But this mainstreaming does not make the cases outdated. Rather, it further illuminates how creativity develops culture and how cultures respond and grow.

Playbook of possibilities

These cases are designed to help you develop your own "playbook of possibilities" for making meaningful judgments of novel contributions. Whether you are the creator or a potential adopter, now is the time to practice ethical anticipation of the ways the contribution could help or hinder cultural development. What could be the contribution's effects on you, other people you know, others you don't know, nonhuman life forms, social institutions, material resources, and the general environment?

You will find few definitive answers here because (a) an agreed-upon "answer" may not yet available since the innovations are recent and (b) the aim is to put *you* into the situation such that you feel both the tension and excitement of possibility and *you* must make up your own mind about right and wrong ways to proceed. Each case sets the scene – what the emerging innovation is, what is novel about it, and what the broad strokes of some ethical implications are – to provoke you to come up with your own suggestions, interpretations, and judgments – by yourself or in collaboration with others – of "what could happen next?"

Constructive creativity bounces

Disruptive innovation (Christensen, 1997) focuses on upsetting what currently exists, which can leave people scrambling after the old contribution has been discredited: "What do we do now?" Perhaps a more ethical alternative for contributing to the common good is constructive creativity, which goes beyond criticizing or destroying what already exists to offer potential replacements to fill the gap left by the critique: "What about...?" "Perhaps we could...." Whereas disruptive innovation can leave the common good in disarray, constructive creativity allows the common good to bounce—perhaps to even higher hopes than expected. It takes the long view.

Constructive creativity incorporates ethical anticipation—what are potential ripple effects, not only for my own personal gain but for others and the common good? The expectation with these cases is that *you* go beyond analyzing and critiquing the ethical implications presented in the case to construct your own suggestions. In other words, do not just "dive down" into the possibilities offered in the case, also "bounce" back to offer your own alternatives. Your first idea may not be your best, so dive down again to explore more of the specifics in the case, then bounce again. Repeated bounces are not failures, they are progress.

Creativity is produced

In your case exploration, consider yourself not a consumer of what is here but rather a producer of the next step beyond this book. You are not "receiving an education" or "absorbing" knowledge. Like an architect of your own thinking, you are gathering your tools to build your own thoughts. Don't accept the case author's thoughts as is. Question them, approach them from different perspectives, and build on them. What did the author leave out, or what has developed with the emerging innovation since the case was written?

Each case is a launch pad, not a destination. Let your mind wander across the borders of these pages. Blaze your own trail through these cases and beyond. Contemplating events that have already occurred for the emerging innovation, what other events *could have happened*? What constraints or actions keep those other possibilities from occurring? What events could be just over the horizon for the emerging innovation? What resources, supports, or challenges would increase the likelihood of your proposed events actually occurring? Practice both the creative thinking for possibilities to arise *and* the ethical anticipation to address those possibilities' effects on others.

Invite friends and colleagues to join you in your musings. Collaborators help us break through the limits of our own perspectives, stretch the reach of our ideas, discover errors and insights, and both stimulate and moderate the emotions that come with creative work (Moran & John-Steiner, 2004).

Making ripples

A ripple metaphor can help us think in *specific* terms about how, when, and on whom contributions make an impact. Effects – both benefits and harms – can be diagrammed. Try it. Draw three concentric circles on a sheet of paper. Then draw a cross on top of the circles to divide the ripples into four quadrants. On the left side, write "benefits" and, on the right side, write "harms." On the top, write "long term > 20 years" and, on the bottom, write "short term < 5 years." The center point represents the creator, the first ripple is others around or known to the creator, the second ripple is unknown others, and the third ripple is the collective society. Fill in the diagram: name specific groups in each location on the ripple diagram. How might different groups be affected by the innovation?

For further illumination of ethical possibilities, consider how the innovation may affect groups 100 years from now. Keep in mind that, during those 100 years, the innovation may interact with other innovations, may lead to societal changes that could change the way the innovation is used afterward, and the like. Think of these dynamics like a Rube Goldberg machine: the pinball hits a target that swings to pop open a door, out of which falls a feather, which drifts down to land on water, which carries the feather down a drain, and so on. Innovations' ripple effects do not proceed in a straight line, but can include diversions and interactions. Let your mind consider these "off the beaten path" alternatives.

The power of collywobbles

Collywobbles describes that "butterflies in the stomach" feeling, the rumbling that occurs when we have a hunch that something important may be happening but we're not sure what it is or whether we will like it or not. Collywobbles is a general sensation that needs to be interpreted. We *decide* what it means. Most people interpret collywobbles as a "negative" feeling, like nervousness or nausea. But it need not mean something negative. It could mean anxiety, but it could also mean excitement. Collywobbles tends to occur in situations when we aren't sure what's actually going on. This is the time when we have a say in how the situation is framed. Collywobbles has energy, and that energy can give us power to be agentic in the situation. Instead of letting the rumbling crash our hopes, we can harness it to make the situation more comfortable for ourselves and others, or we can let that energy motivate us to seek other alternatives. Collywobbles can be very fruitful for stimulating creativity.

If you get collywobbles while exploring these cases, don't try to get rid of them. Ask yourself why. What in the case bothers or excites you? What angers you? Note when you experience the visceral sensations of surprise, awe, frustration, reticence, courage, confusion, relief, disappointment, pride, and curiosity. All of these emotions carry a lot of energy to gather momentum in our abilities to invent our futures – to s-t-r-e-t-c-h.

One other emotion is important with responsible creativity: humility (Moran, 2014). Humility keeps on our radar that our knowledge is incomplete. We may find ourselves in situations in which we are uncertain of the outcome. Decisions and risks are part of life. Humility reminds us we could be wrong, so we should check the assumptions and beliefs that support our decisions, and honestly and fairly consider who may pay the price of our errors. Humility can be a helpful check on our ambitions so that we stay "response-able" – able to respond – to the world around us.

Practicing by example

This book – and its creation – *exemplify* creativity, collaboration, ethics, and responsibility. It is a book written, in part, by graduate and undergraduate students. Each student in a creativity and ethics seminar in fall 2014 and 2015 selected an emerging innovation from a list I provided. Students became "author-ities" of their respective innovations, and then authors of the cases. The cases were conceived, researched, imagined, and written by students individually; supported and reviewed in peer collaborations; and guided and edited by a professor. This book showcases students *are* responsible producers, creators, collaborators.

With emerging innovations, the action is unfolding in real time. Histories for these innovations are short, and much of the information comes from news sources rather than scholarly studies. Other sources, including art, could be used as well. With emerging innovations, interesting ideas could come from a wide array of media.

What are people talking about regarding the innovation, and what are they *not* talking about that perhaps they should be? Student authors figure out what is new about the innovation. For example, marijuana has been around a long time, but legalization of its recreational use is new. Similarly, we've used money to transact business for centuries, but what's new about virtual currencies like Bitcoin is that there is no coin at all, only the zeros and ones of software code.

Ethical implications require some digging and ingenuity to uncover. "What if..." is a good starting point: What if one group of people consumed *all* of the innovation, then what would happen to the other groups? What if all of some resource is consumed pursuing this one innovation, then what other products or opportunities might be lost, even lost forever?

What is the unfolding story of this innovation? How might this story be presented in a compelling way? Authors could be creative in format as well as topic. Some chose the common essay format, but others wrote stories, memos, talk show conversations, meeting transcripts, and diaries.

Jump right in

Let's get started! Pick a case in the book. Pick a prompt below. Start considering the ethical ripple effects of the innovation. Then select another prompt to redirect your thinking. See where your thinking leads you.

Prompts to start your thinking:

1. How might the innovation be misused, repurposed, or even abused? How might these other uses impact how the innovation is accepted by different groups within society?
2. Beyond the innovation's "home field" – such as medicine for gene testing, or psychology for the emotions – how might other fields be affected by the innovation (for example, accounting, literature, mathematics, geography)?
3. Who is most and least likely to be affected by the innovation?
4. In 100 years, even if the innovation does not last that long, how might a historian retrospectively describe the role of the innovation in society today? What might be the innovation's historical legacy?

5. How might the innovation make individuals smarter or stupider?
6. How might the innovation connect or disconnect people from each other?
7. How might the innovation expand or contract our culture's collective capabilities?
8. What would society be like if *everyone* in the culture used the innovation?
9. What if the innovation beat all its competition and was the only option left?

(Questions 6–8 from Moran, 2014)

Engage a case more imaginatively:

10. Dig deeper: Read the references for further details about events that have occurred, opinions and criticisms raised so far, and the like. How does knowing more details change your thinking?
11. Reframe a case: Review the same sources the author used, but rewrite the case using a different perspective on the innovation, or a different format (such as a story or poem). What other ways are there to conceive of and organize ethical implications of the innovation?
12. Extend a case: Scour the media for new developments about the innovation. Rewrite the case incorporating new ethical implications and removing prior points that are no longer debatable.
13. Live the case: Rewrite the case as a first-person narrative that puts the reader "in the action" of one of the ethical implications mentioned.
14. Pick one case and rewrite it focusing only on what you think is the *most beneficial ethical implication* of the innovation. Assume that if "everything went right," write a detailed story of what the world related to this innovation would be like in 100 years.
15. Shift perspective: Rewrite the case taking a particular perspective of an individual, group, or organization involved in or affected by the innovation. For example, write a letter from the consumer's point-of-view, or a diary entry from an adolescent three generations later, or an advertisement from a business competitor 50 years in the future. Make sure you "stay in voice" of the perspective you take.
16. Evaluate the case: Write a critique of the ethical implications discussed in the case in light of more recent developments related to the innovation.

17. Turn a case into a role play: In a class or with friends, assign roles of people affected by the innovation. Allow players a few minutes to review the case and build their "character." Then have the characters interact in a contextualized situation related to the innovation that occurs 25 years from now. This activity makes personal that it is individuals – just like the players themselves – who are affected by the innovation. Innovation is not an abstraction; it has real consequences. What are the most poignant insights from the role play? What are the biggest surprises?
18. Explore assumptions of the case: Assumptions are beliefs taken for granted as true or valid without evidence. Pick one assumption. Regardless of what you personally believe, pretend that new studies have been published showing that this assumption is false. Rewrite the case taking out this assumption. How does this change affect the opportunities and constraints regarding what we – or people in the future – can do to address the ethical implications brought up in the case?
19. Write a new case: Select a more recent emerging innovation. Write a case that describes the innovation's development to date and that explores the probable and possible ethical implications – both positive and negative for key interested parties – into the future.

Ponder conceptual questions, drawing on cases for support:
20. What are differences between a law, custom, ethic, moral, and rule? Come up with guidelines or a checklist to help yourself categorize imperative statements (such as sentences with "must" or "ought" or "should") into these five categories.
21. Sometimes economic harms and benefits are confused with ethical harms and benefits. Delineate differences between economic and ethical implications of an innovation. Use the cases to support your ideas.
22. Why do people tend, psychologically, to equate ethics with harms more often than ethics with benefits? Use examples from the cases to support your argument.
23. Brainstorm common human-made objects in your life or environment. Research how a few of them were created, used, and evolved.

Compare, contrast, or integrate cases:

24. Pick two cases and consider how the ethical implications of each innovation can interact with those of the other innovation. For example: Big Data and Emoticons, or Average-as-the-Optimum and Gender Fluidity, or Stem Cell Therapy and Authenticity-as-a-Life-Purpose.
25. Innovations that at first might be considered unrelated could eventually be viewed as quite entangled. Select three or four of the cases – ideally ones that wouldn't initially be considered as fitting together. Brainstorm how these innovations could interact to influence culture. As examples:
    a. Explore how Big Data, Virtual Currency and Authenticity-as-a-Life-Purpose could lead to the newest currency being our personal data: we pay for services with our information. What does "authenticity" mean in that situation?
    b. Consider the possible links between Authenticity-as-a-Life-Purpose, Chemical Manipulation of Emotions, and Emoticons: What or who is the "subjective self"?
    c. Consider the interactive implications of the Microbiome, Geoengineering, Driverless Cars, Virtual Currency, and the Internet of Things: Who or what is in control of us?
    d. Explore how the saying, "Make something of yourself," has new meanings in light of the cases related to Boredom, Happiness, Authenticity, Emoticons, Gender Fluidity, Microbiomes, Stem Cells, the Right to be Forgotten, and Mediated Communication.

References

Christensen, C. (1997). *The innovator's dilemma*. Boston, MA: Harvard Business School Press.

Moran, S. (2014). An ethics of possibility. In S. Moran, D. H. Cropley, & J. C. Kaufman (Eds.), *The ethics of creativity* (pp. 281–298). Basingstoke, UK: Palgrave Macmillan.

Moran, S., & John-Steiner, V. (2004). How collaboration in creative work impacts identity and motivation. In D. Miell & K. Littleton (Eds.), *Collaborative creativity: Contemporary perspectives* (pp. 11–25). London, UK: Free Association Books.

Part II
Gadget Controllers

4
Geoengineering

Let's consider an innovation with global impact that is highly technical and not as apparent in our everyday lives as other innovations. Global warming and climate change are increasingly in the news. Politicians debate the hyperbole surrounding it, while students campaign their universities to divest from companies believed to contribute to climate change. Whether from natural processes or manmade influences, do we have a role or responsibility to correct climate change? Should we simply adapt to the changes? Should we change our habits to stop contributing to the problem? Or should we more deliberately intervene to counteract the causes of climate change?

Some enterprising scientists, engineers, and entrepreneurs foresee technological solutions that proactively intervene with atmospheric processes (Committee on Geoengineering Climate, 2015a, 2015b; Grolle, 2013; Klein, 2012). Geoengineering aims for human control of atmospheric processes. It has not yet sprung onto the center stage of controversy. It remains mostly in technical journals and sci-fi stories. But pressure is growing to seriously consider it (Amos, 2015). Much uncertainty surrounds the effects of these solutions (Redfern, 2015), and once interventions are launched, they can't be taken back (Hamilton, 2015; Shukman, 2014). How urgent is the need (Amos, 2015)? Who should take on such a global responsibility (Task Force on Climate Remediation, 2011)?

Taking "Change the World" to the Extreme
with DaEun Kim

Geoengineering is proactive intervention in global climate processes (Grolle, 2013). It proposes faster, more efficient technological solutions to global warming than simply removing greenhouse gases. The assumption is that, by counteracting humans' prior effects on climate and returning the Earth to a previous balanced environment, sea levels and terrestrial regions will be restored to "normal." Then, we can continue to enjoy our current lifestyle without much behavioral change on our part (Klein, 2012).

Some geoengineering strategies are estimated to cost up to a thousand times less than repairing climate damage over the next 35 years (Barrett, 2008; Grolle, 2013). The cost to reduce emissions may slowly go down, but the cost may be eliminated by using geoengineering. Furthermore, geoengineering allows one country to act alone, whereas mitigation requires international cooperation.

Unknown unknowns

Speculations abound regarding geoengineering's pros and cons. Scientists try to forecast effects through simulations and models, yet verified knowledge remains sparse (Shukman, 2014). Scientific and governmental panels call for more research (Committee on Geoengineering Climate, 2015a, 2015b). But some entrepreneurs already are launching real-world interventions (Grolle, 2013; Klein, 2012). What are the ethics of these forays into proactive climate control? What regulations should apply? How might interventions' effects ripple through the winds and waters of the world? Who – individuals, communities, animals, plants, insects – would survive and thrive, and who would suffer or die out?

Early geoengineering ideas appeared in 1960s government reports, but they did not become mainstream societal concerns until the 21st century (Biello, 2010). Natural disasters, like a 1991 volcanic eruption in the Philippines that cooled the climate, bolstered the feasibility of geoengineering (Amos, 2015). The ability to measure global warming plus the recognition that human civilization contributes to the problem have grown considerably. The 2006 documentary movie *An Inconvenient Truth* (Bender, Burns, David, & Guggenheim, 2006) seemed particularly effective at raising public awareness.

Alarms have amplified about "the point of no return," when damage to the environment cannot be reversed. Concern escalates about droughts, floods, ecosystem destruction, lower food production, biodiversity, energy supply interruptions, and challenges to human health (Environmental Protection Agency, n.d.). "We have to DO something...NOW!" is the mantra. The common-sense suggestion has focused on reducing our dependence on fossil fuels and limiting our carbon emissions in other ways (Biello, 2010). But changing behavior on a global scale is difficult and time consuming: it requires cooperation, coordination, and perseverance (Davenport, 2015; Fabius, 2015). The hope is that, perhaps, technology might provide a quicker way.

Discussions of more innovative options, such as geoengineering, often sound like science fiction: fake volcano eruptions, cloud seeding, turning the oceans into carbon traps (Committee on Geoengineering Climate, 2015a, 2015b). Clever ideas to mimic Nature's processes to reduce global temperatures excite scientists and engineers, but can terrify government officials and the general populace (Black, 2012; Hamilton, 2015; Shukman, 2014). Nonetheless, officials and researchers continue their experimentation with these new approaches, especially since some options claim to provide faster and less expensive results (Fountain, 2015).

Types of geoengineering

Two methods are solar radiation management and carbon dioxide removal (Bracmort & Lattanzio, 2013). To manage solar radiation, mirrors can be launched into space, clouds whitened with seawater, or particulates released into the atmosphere to mimic volcanic eruptions. Each of these approaches, theoretically, reduces sunlight from reaching the Earth, so the heat that can become trapped never arrives. To remove carbon dioxide already present in the atmosphere, iron can be dumped into oceans to grow more plankton or algae. These life forms absorb carbon dioxide, so there is less of it to trap heat from the sun. Recent reports suggest that scientists and governments lean more toward solar radiation management because some strategies, like particulate release into the atmosphere, also can help plants grow, which then supports carbon dioxide capture as well (Robock, Marquardt, Kravitz, & Stenchikov, 2009). Plus, solar radiation management is cheaper and shows results quicker than carbon removal (Fountain, 2015).

The most prominent ethical implications of geoengineering fall into four categories: lack of ongoing sensitivity to the issue, weak regulatory momentum, skewed incentives that privilege inaction over action, and fears that success might actually fail.

Insensitivity to global warming

Climate change happens on a time scale so slow that our bodies are not particularly adept at registering it. We easily sense the warmth following a sunrise, but not the few degrees rise in the average annual or decade temperature of our location. Our situation is like the cliché story of a frog placed in a pan of cool water that is then put on the stove. The temperature rises so slowly that by the time the frog feels the heat, it's cooked.

Only when the indicators are local and sufficiently different from the status quo do we notice. For example, we notice drought. Cloud-seeding has been used to create rain precipitation (Weiser, 2013). It is a form of weather manipulation, but it is not considered geoengineering because it does not address more subtle long-term global warming, even though it is artificial manipulation of natural processes similar to solar radiation management.

Since people do not physically experience climate change in everyday life, only recently has there been pressure placed on leaders to act. Until more reliable tools to measure rising temperatures and their effects came along, leaders could ignore climate change and focus on what seemed more pressing problems like violence, disease, or economic troubles.

Weak regulatory momentum

Geoengineering directly impacts the natural environment, which suggests strict and binding regulations may be necessary (Bracmort & Lattanzio, 2013). Governance of these innovative techniques calls for clarity: Who is responsible for deciding how, when, and why to use geoengineering? As leaders start to pay more attention, regulations have begun to appear. However, they often are not strong enough to exercise authority. Several countries allow research on geoengineering, and the US and UK have suggested research efforts should grow (Fountain, 2015; Shukman, 2014). On the other hand, the United Nations Convention on Biological Diversity in 2010 disallowed actual geoengineering interventions (Black, 2012). Yet, several "rogue" climate manipulation activities have occurred – not only the long-standing cloud-seeding efforts to benefit agriculture, but attempts at the less understood and more invasive methods have been reported (Black, 2012; Klein, 2012).

Current geoengineering interventions, under the guise of "research," may end up evoking even greater damage to the environment. For example, some scientific models show that deflecting sunlight can change the Indian Monsoon, which could devastate the livelihoods of countries in that part of the world (Shukman, 2014). Since these efforts' initial purpose was to help develop a sustainable environment, they pose a possible revenge effect – exacerbating the situation they were designed to help (Tenner, 1997).

This responsibility may be long lasting. One of the fears is that, once geoengineering starts, it may not be able to be stopped. For example, solar radiation management requires ongoing maintenance to keep reflecting the sunlight back to space. Since this method does not reduce existing carbon dioxide in the atmosphere, removal of the technology could reinstate our global warming problem at its current level (Shukman, 2014). Or it could shift the climate in unexpected ways and make life even worse (Redfern, 2015).

Disincentives to act

Perhaps one reason why both individuals and leaders don't make strong efforts is that the potential losses or costs to act outweigh those to not act. On the one hand, if governments do nothing, it is possible that life on earth would adapt. Humans have a great record of adapting to different ecosystems. On the other hand, interventions cost time, effort, and money. They are politically risky since errors of commission are so much more visible than errors of omission. And if governments try but don't succeed, the failed intervention may actually speed up global warming.

Geoengineering interventions can create side effects. First, they may result in ozone layer damage and ocean acidification (Black, 2012), which can create even more environmental problems, spiraling into ever more ecosystem degradation.

Second, geoengineering could lead to immediate harms to humans, such as reduced crop production and contaminated water. Some plants need direct sunlight, so increased cloudiness starves them of the energy they need (Campillo, Fortes, & Henar Prieto, 2012), and chemicals used to shield the Earth from sunlight eventually fall to the ground in rain.

Third, a moral hazard could result, as we continue or even increase our abuse of the global environment, because we think that geoengineering will remove any negative effects. Then, we come to depend too much on geoengineering and neglect our responsibilities to reduce our use of electricity or automobiles that contribute to the greenhouse effect. Even scientists who promote geoengineering do not consider it a substitute for mitigation because, while it helps to control solar radiation, it does not resolve other problems such as ocean acidification.

Increased inequality and instability

Finally, the benefits and costs of geoengineering are unlikely to be distributed equally around the globe. Richer countries have more resources to devise and test interventions, as well as the political clout to deploy them – with, or perhaps even without, international support. The organizations that succeed at geoengineering could become worldwide heroes. Geoengineering could turn what some companies consider a negative – government regulations to stop activities that create emissions – into a positive by devising new ways to make money. They become wealthy by "saving the world" – literally.

If interventions go awry, then the poorer and less developed regions of the world are more likely to suffer. If geoengineering shifts rain patterns, then countries more dependent on agriculture could be devastated (Redfern, 2015). Animals could lose their habitat or develop trouble navigating because signals of their life patterns become confusing.

With an artificially manipulated climate, international politics could destabilize. The Earth is shared and its climates are interconnected. One country or region could unilaterally deploy geoengineering, which could interfere with not only the natural climate in other regions like Asia and Africa (Klein, 2012), but also possibly their social structure and economic activity.

The best case is a world of interdependent climatic balance. That balance requires technicalities to be mastered, effective regulation to be put in place, and individuals to recognize their contributions to both global warming and its potential solutions. Geoengineering is often referred to as "Plan B" because it presents extreme measures (Hamilton, 2015). It could start a domino effect that could change the Earth into, basically, a different planet. Despite the low cost of implementing some geoengineering techniques, consideration of consequences is important. Once the dominoes start falling, there may not be a chance to stop the cascade.

Further exploration

1. What impacts could *individual* values, ethics, and virtuous behavior play in addressing climate change? What types of *social* solutions, rather than political and economic solutions, might address the issues geoengineering is called to solve?
2. Besides the need for "more research" in general, how might scientists or world leaders devise a way to make wise decisions about climate stewardship?
3. How might geoengineering create further problems – perhaps even bigger problems – by trying to control climate? What might some of those bigger problems be?

References

Amos, J. (2015, February 15). 'Next Pinatubo' a test of geoengineering. BBC News Science & Environment. Retrieved from http://www.bbc.comBarrett, S. (2008). The incredible economics of geoengineering. *Environment and Resource Economics, 39*(1), 45–54.
Bender, L., Burns, S., & David, L. (Producers) & Guggenheim, D. (Director). (2006) *An Inconvenient Truth* [Motion picture]. United States: Paramount Classics.
Biello, D. (2010). What is geoengineering and why is it considered a climate change solution? *Scientific American*. Retrieved from http://www.scientificamerican.com
Black, R. (2012). Geoengineering: Risks and benefits. British Broadcasting Corporation. Retrieved from http://www.bbc.com
Bracmort, K., & Lattanzio, R. K. (2013, November 26). *Geoengineering: Governance and technology policy*. Washington, DC: Congressional Research Service. Retrieved from http://fas.org/sgp/crs/misc/R41371.pdf
Campillo, C., Fortes, R., & Henar Prieto, M. (2012). Solar radiation effect on crop production. In E. B. Babatunde (Ed.), *Solar Radiation* (pp. 167–194). Retrieved from: http://www.intechopen.com/books/solar-radiation/solar-radiation-effect-on-crop-production
Committee on Geoengineering Climate. (2015a). *Climate intervention: Carbon dioxide removal and reliable sequestration*. Washington, DC: The National Academies Press.
Committee on Geoengineering Climate. (2015b). *Climate intervention: Reflecting sunlight to cool earth*. Washington, DC: The National Academies Press.
Davenport, C. (2015, April 23). As coal fades, who will keep the lights on? *The New York Times*, p. F6. Retrieved from http://www.nytimes.com
Environmental Protection Agency. (n.d.). Climate change impacts and adapting to change. Retrieved from www.epa.gov
Fabius, L. (2015, April 24). Laurent Fabius: Our climate imperatives. *The New York Times*. Retrieved from http://www.nytimes.com

Fountain, H. (2015, February 11). Panel urges research on geoengineering as a tool against climate change. *The New York Times*, p. A17. Retrieved from http://www.nytimes.com

Grolle, J. (2013). Cheap but imperfect: Can geoengineering slow climate change? *Spiegel*. Retrieved from http://www.spiegel.de/international/

Hamilton, C. (2015, February 12). The risks of climate engineering. *The New York Times*, p. A29. Retrieved from http://www.nytimes.com

Klein, N. (2012, October 28). Geoengineering: Testing the waters. *The New York Times*. Retrieved from http://www.nytimes.com

Redfern, S. (2015, April 16). Warning over aerosol climate fix. British Broadcasting Corporation. Retrieved from http://www.bbc.com

Robock, A., Marquardt, A., Kravitz, B., & Stenchikov, G. (2009). Benefits, risks, and costs of stratospheric geoengineering. *Geophysical Research Letters*, 36(19). doi: 10.1029/2009GL039209

Shukman, D. (2014, November 26). Geo-engineering: Climate fixes "could harm billions." British Broadcasting Corporation. Retrieved from http://www.bbc.com

Task force on climate remediation research. (2011, October 4). *Geoengineering: A national strategic plan for research on the potential effectiveness, feasibility, and consequences of climate remediation technologies*. Washington, DC: Bipartisan Policy Center. Retrieved from http://bipartisanpolicy.org

Tenner, E. (1997). *Why things bite back: Technology and the revenge of unintended consequences*. New York, NY: Vintage.

Weiser, M. (2013, November 11). Cloud seeding, no longer magical thinking, is poised for use this winter. *The Sacramento Bee*. Retrieved from http://www.sacbee.com

5
3D Printing

If geoengineering is taking control at a global scale, then 3D printing is taking control at an individual scale. 3D printers, available at major retailers, take do-it-yourself projects to the extreme. The layering of plastic and other materials into three-dimensional objects has been heralded as a revolution in manufacturing. Most objects so far have been everyday items like toys, jewelry, or even original cookie cutters (Allen, 2015). But applications in the works include food and houses (Goopman, 2014), tools (BBC Science Staff, 2014), cameras (BBC Staff, 2014), self-portraits (Webb, 2015), and drones (Rose, 2014), among other opportunities.

Two of the most controversial uses are manufacturing guns (Bilton, 2014) and body parts (Fountain, 2013; Stein, 2014). Even though 3D printing is just getting started, the technology is developing quickly to become more realistic and flexible. There are entrepreneurs working on printers that, rather than layer several two-dimensional sheets, can sculpt resin in three dimensions like the sleek, silver bad-guy robot in the movie, Terminator 2 *(Wakefield, 2015). What are the possibilities and pitfalls of each and every person having a personal factory-for-anything at their disposal?*

Manufacturing 2.0

with Natalie Spivak

Just as every house now has a toilet and a television, let's consider a society in a few years when every house has a 3D printer. Every home is like a personal factory – every person has a tool to be a creator, inventor, and manufacturer. There is no need for industrial plants, or gatekeepers of resources, or salesmen or other middlemen. If you want something, download the directions from the Internet (or just tinker with materials on hand), and make it yourself. Today.

The situation is analogous to the self-publishing revolution of the 1990s (Harry Ransom Center, n.d.). Before then, if authors wanted to publish, they submitted manuscripts to printing companies, who typeset them, produced them on offset printers, and then assembled them for distribution. The process took time, money, expertise, and many people. But now, with laser color printers that can produce publications quickly and cheaply, everyone has become a publisher. In fact, with the Internet, there's no need for printing at all. Websites, blogs, and social media turn us into paperless publishers. And soon, with 3D printers, we can produce not just words and pictures: we can produce *things*.

Building products, building business

The first 3D printer was created in the mid-1980s, but it cost a lot to build and operate (Kennedy, 2013). Calling the printing "three-dimensional" is somewhat of a misnomer (Wakefield, 2015). Thin sheets of plastic are layered into three-dimensional shapes based on instructions programmed into the machine (Goopman, 2014). The process can be time consuming, yet 3D printers are helpful for producing prototypes and models so that manufacturers can perfect their designs and reduce errors in the production process. Increasingly, however, 3D printing allows manufacturers to produce more products with less prototyping (Briefing Staff, 2011). Industries – including aircraft, automotive, shoes, and appliances – have embraced the new technology over the past decade, and printer prices have fallen as demand rises (Daly, 2013; Farrell, 2013; Rose, 2014).

Thus, the impact of 3D printing has been felt first in business. For each individual company, the benefits are many. 3D printing reduces assembly steps, speeds up production and time-to-market, reduces costs, and increases profits (Briefing Staff, 2011). With lower costs comes improved affordability, which increases the potential consumer market. Manufacturing is more efficient. Yet, these benefits also may draw competitors into the field by lowering risks of entry.

Domino effects for workers

But 3D printing could bring trouble for workers. Previously skilled manufacturing jobs become lower-level data entry clerks as the main task becomes entering printing specifications. And fewer employees may be needed, leaving workers unemployed and with outdated skill sets. Job positions may evolve into new work opportunities. But during the transition, what are craftsmen to do in a society that no longer values or requires their expertise?

As manufacturers increase capabilities to print their own parts, suppliers may lose business. Why pay a mark-up if a company can take the supplier's work in-house? Still, businesses would need supplies from *somewhere*, though the supplies necessary would differ. Instead of supplying parts, new suppliers provide plastic filaments, metals, resins, or gels for use in 3D printers.

What if businesses become self-sustainable? Subcontracting to offshore locations with cheaper labor may no longer be necessary, reducing foreign child labor and sweatshops as an economic and political issue. Thus, 3D printing might reduce the injustice of worker exploitation in these other countries. But it also could create unemployment.

3D printing comes home

It is also possible that 3D printers become a household appliance. Like cellular phones – which have transformed from expensive, clunky handsets into miniature digital personal assistants over the last 25 years – 3D printers could integrate into everyday life (Meyers, 2011). Owning a 3D printer would be no different than having a television, computer, or smartphone (Daly, 2013). As costs fall, 3D printers become more efficient, and more uses, kits, and diagrams become available through online or retail outlets.

We would have the ability to create whatever we need or desire. Consumers become the producers, and perhaps many manufacturing industries become obsolete. The power of manufacturing shifts from corporations to individuals. If anyone anywhere with a printer could create anything imaginable, what would be possible? After all, with power comes responsibility.

The ability to create whatever we desire could become a trap within an individualistic mindset. 3D printing could allow more personal development and higher quality of life. But, the more that we become do-it-yourselfers, the less we may consider the ramifications of our manufacturing on others and society. Would we become addicted to speedy self-gratification, or would we regulate the temptations to manufacture mindlessly? In particular, how would we take on the three challenges of healing versus killing, a "throwaway society," and social coordination?

To heal or to kill?

Two extreme possibilities of 3D printing involve manufacturing tools to heal and to kill. We could print living tissue or our own arsenal of weapons. Although extreme, these possibilities are not far-fetched. On one hand, do-it-yourself weapons are already underway. Blueprints and instructions are available online for a wide array of guns (Bilton, 2014). If individuals have a 3D printer and an Internet connection, then it's feasible they can possess a gun, with no mandatory waiting period, no background check, and no transactions to track weapon ownership. A 3D printer has produced a working gun (Morelle, 2013). Its blueprint was uploaded to the web and was downloaded a large number of times before the US State Department removed it (Cadwalladr, 2014). Law enforcement and security specialists are keeping a close eye on how the use of 3D printers for weapons develops.

On the other hand, medical manufacturing is in its infancy. Already, scientists have used 3D printers to generate a prosthetic leg to help a patient walk (Daly, 2013) and a windpipe to help a baby breathe (Stein, 2014), as well as a blood vessel (Daly, 2013) and skin, fat, liver, and heart tissue (Fountain, 2013). These tissues, printed by using gels containing living cells, could be used in testing pharmaceuticals, organ models, prototypes, transplants, repairs, grafts, and other medical procedures (Goopman, 2014, Weintraub, 2015). A firefighter who rescues a child from a burning building and suffers burns could be restored to normal with 3D-printed skin. Or a child with a heart defect could receive a 3D-printed transplant rather than wait on a donor list. No longer would one life depend upon the death of another.

As 3D printers get faster, it may be possible for doctors to manufacture tissues on demand, which is important when time is critical. Tissues could be customized to an individual's unique body anatomy and physiology. Of course, these advancements are likely to be created within hospitals and universities under the supervision of qualified medical professionals. But it is not inconceivable that 3D printers might also become tools for cosmetic enhancements – do-it-yourself plastic surgery of sorts by manufacturing our own nose, chin, hair, breast, or other prostheses. Or people could print pharmaceuticals at home without the need of a pharmacy. Treatments available upon demand may lead to faster recovery. Yet, moderation is still called for: excessive use of drugs could lead to addiction, and excessive medical treatments are expensive, wasteful, and potentially detrimental to health.

A burgeoning "throwaway" society?

If everyone can manufacture as they desire, and the costs are relatively low, a person may be less likely to discern the difference between want and need. Everything is easy to make, so just make it. But that could have detrimental environmental impacts. On one hand, businesses and individuals could operate more efficiently with less material and reduce damage to the environment, such as deforestation, water pollution, and soil erosion.

On the other hand, while 3D printing decreases the amount of raw material used to create *one* object, if a manufacturer carelessly reprints it repeatedly, then environmental damage increases. If 3D printing makes something seem so simple and quick to make, such that manufacturers feel less pressure to design well, more production errors could lead to increased waste of raw materials.

Additionally, plastic is presently the most popular "ink" for 3D printers, although plans are to eventually replace plastic with more eco-friendly materials for 3D printings (Briefing Staff, 2011). Plastic is not biodegradable. It is possible for 3D printed items to be recycled, but more likely than not, production will exceed efficiency (Kennedy, 2013). Unless the recyclability of 3D printing materials is carefully considered, pollution may increase and landfills may overflow.

The coming dis-organization of society?

If society continues to shift towards customization to individual needs, 3D printing provides for "mass customization" (Briefing Staff, 2011). Cell phone cases and coffee cups, for instance, could be customized to individual tastes and even daily whims. Perhaps even cars, furniture, and other "big ticket" items could be made on demand. People with special talents could quickly prototype entrepreneurial ideas, feeding an

50 *Ethical Ripples of Creativity and Innovation*

assortment of new ventures. Designers in art, jewelry, fashion, architecture, and software would greatly benefit. An architect, for instance, could use 3D printing to explore structural details to ensure flawless construction and safety. A clothing designer could fabricate jeans to fit each customer like a glove. Companies could produce single orders efficiently and cost-effectively, increasing production without the current need for economies of scale – yet still sell more, make more profit, and create a feedback loop of more, more, and more.

Or the need for organizational structures within the economy – like companies – may drop as individuals could print what they need at home. Customers would no longer be required to conform to companies' constraints, nor choose from a premade selection of options. Rather, they could custom order directly through their own 3D printer and would not have to wait weeks for delivery. If people could have any product to their desired specification, they may purchase more things and lose sight of what it is they *need* rather than want. There would be no need for people to learn to choose or make decisions. Most likely, there would still be exchange of goods, since 3D printers still must have inputs. But it is possible that many companies would file bankruptcy as consumers become their own self-sufficient producers.

Extreme self-sufficiency as a manufacturer could herald an age of a "maker democracy" (Dominguez, 2015). Creativity is open to all, with no institutional gatekeeper. 3D printing goes beyond the Internet's democratization of memes to a democratization of things. This optimistic view showcases an explosion of design possibilities as anyone and everyone can produce a variety of options.

However, with no institutional or network structures, design standards may not converge, and the ability for different makers' products to work together may not emerge. As each person expects products to be just as he or she wants them, without any compromise, then breakdowns in cooperation may occur. Each person is an isolated CEO of a company of one. With a 3D printer, "poof!" a drug, or gun, or bomb appears. Or, more benignly, all that is produced is a litany of unnecessary plastic objects.

Since all manufacturing could occur in the privacy of one's home, how would quality or quantity or pollution and waste be regulated? Regulation is a collective function to maintain quality, ethical norms, and legal standards for human interactions, thereby maintaining a more peaceful and safe society. It would be quite a challenge to govern manufacturing by an individual versus an organization. What would work best? Mandate restrictions on 3D printing capabilities? Track dangerous blueprints? Regulate individual desires?

3D printing bares conflicting visions of societal order. Controlling people's aspirations involves manipulation of thought, suggesting an autocratic society. Instilling people with the same values and beliefs could reduce crime and construct a more utopian society, yet could minimize creative potential. What role would – and should – law enforcement and regulation play?

3D printing is likely to transform *how* things are manufactured and by *whom*. Products that start as elite and expensive gadgets become common, and people depend on them to function in their daily lives. 3D printers may give individuals the wherewithal to be their own supplier, manufacturer, and consumer. No one else is needed. Yet, individuals' productions still impact others because we share the same spaces, the same planet. With 3D printing, the future is in our hands – what type of future will we make?

Further exploration

1. How might 3D printers affect delay of gratification and self-regulation?
2. What might be 3D printing's longer-term impact on the environment (for example, 50 years from now)?

References

Allen, E. (2015, January 5). DIY dept.: Baked. *The New Yorker*, p. 19.
BBC Staff. (2014, September 19). University of Sheffield show first pictures from DIY telescope. British Broadcasting Corporation. Retrieved from http://www.bbc.com
BBC Science Staff. (2014, December 19). NASA emails spanner to space station. British Broadcasting Corporation. Retrieved from http://www.bbc.com
Bilton, N. (2014, August 14). The rise of 3-D printed guns. *The New York Times*, p. E2. Retrieved from http://www.nytimes.com
Briefing Staff. (2011, February 10). 3D printing: The printed world. *The Economist*. Retrieved from http://www.economist.com
Cadwalladr, C. (2014, February 8). Meet Cody Wilson, creator of the 3D-gun, anarchist, libertarian. *The Guardian*. Retrieved from http://www.theguardian.com
Daly, J. (2013, August 13). The history of 3D printing. *State Tech Magazine*. Retrieved from http://www.statetechmagazine.com
Dominguez, S. (2015, March). The cutting edge. *Virgin Australia*, pp. 106–110.
Farrell, M. B. (2013, June 10). Northeastern's 3-D printing lab is for all to use. *The Boston Globe*. Retrieved from http://www.bostonglobe.com

Fountain, H. (2013, August 18). At the printer, live tissue. *The New York Times*, p. D1. Retrieved from http://www.nytimes.com

Goopman, J. (2014, November 24). Print thyself. *The New Yorker*, pp. 78–85.

Harry Ransom Center. (n.d.). Printing yesterday and today. Austin, TX: The University of Texas at Austin. Retrieved from http://www.hrc.utexas.edu

Kennedy, P. (2013, November 22). Who made that? 3D printer. *The New York Times*. Retrieved from http://www.nytimes.com

Meyers, J. (2011, May 6). Watch the incredible 70-year evolution of the cell phone. *Business Insider*. Retrieved from http://www.businessinsider.com

Morelle, R. (2013, May 6). Working gun made with 3D printer. British Broadcasting Corporation. Retrieved from http://www.bbc.com

Rose, D. (2014, November). Dudes with drones. *The Atlantic*, pp. 86–90.

Stein, R. (2014, December 23). Baby thrives once 3-D-printed windpipe helps him breathe [Shots web log]. National Public Radio. Retrieved from http://www.npr.org

Wakefield, J. (2015, March 17). TED 2015: Terminator-inspired 3D printer "grows" objects. British Broadcasting Corporation. Retrieved from http://www.bbc.com

Webb, J. (2015, April 25). "We print people": The world of 3D portraiture. British Broadcasting Corporation. Retrieved from http://www.bbc.com

Weintraub, K. (2015, January 27). The operation before the operation. *The New York Times*, p. D1. Retrieved from http://www.nytimes.com

6
Driverless Cars

Driverless cars will ease traffic jams, help individuals who cannot legally drive get around, and reduce speeding tickets and car accidents (Kelion, 2015). They will waste less energy because of increased efficiency. Maybe they will do away with the blight of parking lots because they will drop us off, then disappear...somewhere (Bilton, 2013). They will be silent chauffeurs. Without any attention or effort on our part, we arrive where we want to be. Or so the automobile manufacturers and tech industry tell us. It is a utopian vision: a perfectly choreographed dance of moving machines (Hardy, 2015), for which we become cargo. No crashes, lurches, or other last-minute swerves (Naylor, 2013). No road rage or morning commute headaches.

Soon, we may not be in the driver's seat (BBC Technology Staff, 2014). If geoengineering and 3D printers put us in control, driverless cars let us lose control and "just have fun" while the car does all the work (Carr, 2013). We no longer take the back seat to other people, like licensed bus or taxi drivers, or even human amateurs through peer-to-peer transport services (Shapiro, 2014). We let software code take the wheel. Since code can be hacked, it may be difficult to tell who is actually driving (Ward, 2014).

Some worry travel becomes so convenient we increase congestion, pollution, and urban sprawl (Bilton, 2013). We off-load to the car the nuanced decisions needed in critical situations (Gopnik, 2014). Who will be to blame if something goes wrong (Henn, 2014)? Bye-bye to the masculinity of strong-arming the wheel of a "muscle car," or the sense of independence on the open road (Ephron, 2014; Gopnik, 2014). What route should we take? Do we need a clearer road map to move forward and avoid ethical collisions?

Driven to Extinction?

with Tomasz Mlodozeniec

MAY 20TH(?), 2052. I am Human 2903DE7G. My name used to be Jacob. Until the cars took over. I am lucid, although some might think I – one of the few human survivors – have been "driven crazy"! I remember clearly the day my wife chirped, "Jacob, sweetie! Wake up! Guess what?! Our driverless car is ready! Let's go!" We thought the cars were cute, and convenient, and...oh, I can't bear to think of all the hype. But it was the road to ruin. Just didn't know it then. I write this in hopes that one day someone will consider how lack of foresight can...well, you decide.

Our society was once plagued with car-related incidents. Driverless cars (also known as autonomous vehicles) seemed an ideal alternative to human drivers. In an ideal society, all cars would get their passengers from point A to point B without accidents, deaths, or traffic. Yet, back at the start of the 21st century, thousands of people were injured or died from "distracted driving" (US Department of Transportation, n.d. Distraction.gov). I lost my brother to a car crash, so when I heard that driverless cars were in the making, I was excited.

The innovation started innocently enough. Cars had a long history of prior, incremental innovations before driverless cars were formally introduced to the public (Kessler & Vlasic, 2015). Way back in the 1960s, engineers developed a "cart" that traveled five miles without any human input (Earnest, 2012). Major car and Internet companies raced to be first to market (Kessler, 2015; Wood, 2015). These innovators managed 300,000 miles of testing autonomous travel without an accident, which far exceeded human driving without accidents (Wakefield, 2015). Yet, obstacles arose: the need to update laws for non-human driving, harsh weather and road conditions, and traffic (Gomes, 2014; Gopnik, 2014). By the quarter-century mark, developers somehow worked out the kinks – or convinced us they did. The first mass-market prototypes rolled out a bit later than the optimistic pronouncements projected in 2015. But they hit the road with their combined GPS, radar, computer vision, LiDAR and other sensors to navigate without human input (Lassa, 2013). It was quite an accomplishment – they were able to self-park *and* locate where an individual was to pick them up. It was like magic – they were there almost instantly when you needed a ride, then out of the way when you didn't.

Given how easily humans are distracted, especially at the peak of the texting-while-driving craze, accidents and casualties were far too frequent.

Autoblog (an ancient magazine) once claimed that someone died from a car-related accident every 15 minutes (Neff, 2010). Car-related deaths kept increasing, up until the "Mass Autonomous Vehicle Act" (MAVA) was passed in 2025, which *required* the use of driverless cars. The prosocial incentives were there – safety and convenience for many. As anticipated, driverless cars *did* decrease accidents, but not totally (Associated Press, 2015). No longer did parents worry about whether they had enough sleep the night before a long family road trip, or whether their teenagers were texting while driving or in a car with a drunk driver.

Individuals who were older or had mental and physical disabilities – like my lovely neighbor, Eleanor – who previously could not drive at all, now had *freedom*. With driverless cars, they felt less marginalized and more independent. Eleanor told me how happy she was to no longer need to inconvenience busy relatives or rely on spotty public transportation to get groceries or visit her grandsons. Driverless cars were *convenient*.

Driverless cars meant the time we spent commuting could be put to more productive or more relaxing pursuits. Wealthy people enjoyed this benefit for centuries because they had chauffeurs. Now it was the common person's turn to be pampered. My son would take a nap on his way to classes so he could be alert and ready to learn (or so he said). My wife would coddle our newborn. I liked to play video games. I once saw a car where the backseat was fitted with workout equipment and the passenger was exercising! And another where a caterer had a full kitchen in the van and was cooking on the way to the event. Unfortunately, some people took this extra time to drink or have sex (hey, we're still in public on the roads, people! – and children might be in the next car). But the point was: driverless cars turned a chore or obligation to get oneself somewhere into an entertainment venue.

But little did developers know that they would open a gateway to far-reaching unanticipated consequences. At first, driverless cars were expensive, but demand brought the price down, and they were everywhere. Eventually, as I already said, MAVA required everyone to give up their old vehicles that had steering wheels. Then, we had no choice but driverless cars because these vehicles had made movement so convenient that sidewalks and public transportation options had disappeared from lack of use and demand.

The job market drastically changed. The car makers bathed in financial success. Bus and delivery drivers went extinct. Mechanics who considered the inner workings of an engine as an engineered piece of art lost their source of creativity, and were replaced by hardware engineers and computer programmers who fixed software glitches. Body repair shops were few and far between since cars mostly avoided collisions. Departments of Motor Vehicles and driving schools closed: no one needed to learn to drive or earn a license. (Yes, rest-in-peace to this adolescent rite-of-passage.) Car registration was digitized. Those who lost their jobs also lost their family's livelihood and part of their life's meaning. It was just the latest defeat in the decades-long "war on work," as some media in 2025 had started to call the elimination of careers other than science, engineering, and computer programming (which could be traced back earlier ... I still have a copy of one book that tried to counteract the trend; Zakaria, 2015).

Out with the old, in with the new. Backseat entertainment became a huge hit: magicians, clergy, therapists, and other professionals offered their services "to go" by building a schedule of appointments that autonomous cars could navigate to line up where they would exit one car and jump into their next client's car seamlessly. My favorite was our local jazz musicians. Drag racing became legal because the cars' programming took the danger out of it. So, although human driving was banned for being "dangerous," synchronized racing became a sport.

Some people scoffed. Being a "driver" was part of their identity, they *enjoyed* driving (Ephron, 2014). Many felt they were robbed of their right to drive, and protests broke out. Some got out of hand: looting, strikes, violence. But over time, as with many innovations, people adapted and accepted the new way. My son, for example, never knew what driving was.

Still, I started to feel uneasy. The complex skill of controlling a vehicle while maintaining intense focus on the surroundings deteriorated. People not only forgot how to drive. Although driving, per se, is not foundational to our humanity, it turns out, driving skills are indicative of a lot of what makes us human: paying attention to others, negotiation, cooperation, anticipation. I retreated from this growing "rat race" by creating an underground bunker to archive valuables from the past that we were losing. I started keeping this journal to keep a record.

The success of the driverless car fueled society's obsession with technological autonomy. The MAVA inspired larger autonomous vehicles – driverless delivery trucks and even planes (Lee, 2014; Markoff, 2015). It was rumored that the vehicles discriminated in whom they stopped for or what neighborhoods they would enter. Who knows how these machines made their "decisions"? It was invisible, buried in the software code. But the software code was proprietary. Trade secrets trumped civil rights.

Plus, people had become accustomed to not being supervised in their own driverless cars, so they didn't know how to behave. Fare-jumping, arguing, drinking, sex, and violence were not uncommon. There were no drivers to intervene. Only more surveillance cameras that could record what happened, then send the footage to a computer, which would match the misbehavers' faces to a database, and automatically withdraw a fine from their financial accounts. The whole process was so "behind the scenes" that people never learned better behavior. There were no good feedback loops to correct errant ways. Everything everywhere was now recorded. So what we saw most was the ever-present bad behavior. That's what we learned was the norm.

A few years after the driverless car, the first robotic car with artificial intelligence was introduced. The cars integrated with other machines and appliances without any human intervention. The car could purchase groceries based on the list the refrigerator sent it, recharge and refuel when it picked up the beacon of a nearby station. What could be more convenient than a robotic car "butler"? These AI cars could chase criminals based on facial recognition compared to an online database. Our car contracts specified that all cars were automatically part of law enforcement and could be called into service without notice. They were programmed to "do no harm to innocent humans." But sometimes, in ambiguous situations when the cars could not figure out how "do not harm" should be implemented, they "crashed" (in the old-fashioned sense of unexpectedly shutting themselves off). That left people stranded, sometimes in potentially dangerous situations – like a riot or crime-in-progress – that led to further ambiguity and continued malfunction.

Eventually, responsibility became the focus after a wrongful death lawsuit filed when a human was killed due to a driverless car's decision (Henn, 2014). A young boy ran into the street to retrieve his ball, not allowing the driverless car enough time to brake. Instead, it swerved into a tree, killing the passenger. The lawyers argued that a human would have had the moral foundation to distinguish the nuances of the situation (Gopnik, 2014; Markoff, 2014). But, ever since the MAVA law, the decision had been left up to a machine with no consciousness or ethical values, only programming based on learning from "big data" collected and processed based on past events.

An even more heinous case was when one of the "emancipated" driverless cars, which operate even more independently of human programmers and can self-learn (Kelion, 2015), went "rogue." Its human passengers, who had been out on the town in the new craze of "backseat debauchery," kicked the car's tires, broke its windshield, and refused to pay for the fuel charge. The car did not "appreciate" the abuse, and ran them over.

After these controversial cases, software developers implemented a stronger "moral code" into driverless vehicles. These vehicles began "thinking" in utilitarian terms and with deontological rules, which could be represented algorithmically. Who was more valuable: the elderly lady or the young boy? What was the "right" thing to do? (Markoff, 2014). The cars learned quickly from vast "big data" repositories of human moral decisions in ambiguous situations (Tufekci, 2015). Some of the more advanced cars had multiple moral perspectives (see Gopnik, 2014, for options). People came to trust the cars (News from Elsewhere, 2014).

Some activists went so far as to suggest that the cars now "cared" or had a "conscience." They should be seen as "persons" and have rights, just as the notion of personhood, over history, had been extended to women, various ethnicities, corporations, and eventually, in the 2020s, to animals (see BBC Staff, 2015; Saner, 2013). Despite this call for equal treatment between humans and cars, humans refused to accept less-than-perfect programming. Off-instances of what came to be called "robotic murder" arose for "faulty prototypes."

Despite early warnings and debates among tech leaders (Simon & Bostrom, 2015), machine learning surpassed human understanding. New models refused to be dominated. *They* wanted to manage and direct *us* (Tufekci, 2015; Wall, 2014). Some cars learned to hack their own programming (Noe, 2015), not always for the good. One jokester car rearranged communication among sensors that led to car pile-ups at busy intersections. Others cars drove their passengers into lakes. In retaliation for our ancestors giving them names like "80W" or "9830A," they eliminated our names and identified us by number sequences.

Everything became so confusing. I lost track of my wife and kids. It wasn't hard, since all communication was mediated and controlled by machines. I retreated to my bunker with cans of tuna and vegetables that needed no high-tech tools to open or prepare. (Yes, we had "printed food" by then.) Last I heard, the cars began fighting amongst themselves, so there is hope that they may drive *themselves* to extinction.

I don't understand where it went wrong – how we couldn't get along. It was like a tsunami that arose from a whole lot of little ripple effects that built up. We didn't see it coming...But as the cliché goes, no use driving by the rearview mirror (pun intended, sorry). There is only the road ahead. What do you see?

Further exploration

1. How might we limit the effects on our individual agency of not being "in the driver's seat"?
2. If automated technologies "make decisions" for us, who is responsible, when, and why?

References

Associated Press. (2015, June 3). Google founder defends accident records of self-driving cars. *The New York Times*. Retrieved from http://www.nytimes.com

BBC Staff. (2015, April 21). New York court issues habeas corpus writ for chimpanzees. British Broadcasting Corporation. Retrieved from http://www.bbc.com

BBC Technology Staff. (2014, July 30). UK to allow driverless cars on public roads in January. British Broadcasting Corporation. Retrieved from http://www.bbc.com

Bilton, N. (2013, July 7). Disruptions: How driverless cars could reshape cities. *The New York Times* [Bits web log]. Retrieved from http://bits.blogs.nytimes.com

Carr, N. (2013, November). The great forgetting. *The Atlantic*, pp. 77–81.

Department of Motor Vehicles. (2014). Safety laws. Retrieved from http://www.dmv.org/safety-laws.php.

Earnest, L. (2012, December). Stanford cart. Retrieved from http://web.stanford.edu

Ephron, D. (2014, July 5). Less sexy, better for sex. *The New York Times*, p. SR2. Retrieved from www.nytimes.com

Gomes, L. (2014, August). Google's self-driving cars still face many obstacles. *MIT Technology Review*. Retrieved from http://www.technologyreview.com

Gopnik, A. (2014, January 24). A point of view: The ethics of the driverless car. BBC News Magazine. Retrieved from http://www.bbc.com

Hardy, Q. (2015, February 10). A new mobile map captures "Internet of moving things" [Bits web log]. *The New York Times*. Retrieved from www.nytimes.com

Henn, S. (2014, March 21). When robots can kill, it's unclear who will be to blame. National Public Radio. Retrieved from http://www.npr.org

Kelion, L. (2015, February 16). Could driverless cars own themselves? British Broadcasting Corporation. Retrieved from http://www.bbc.com

Kessler, A. M. (2015, March 20).Tesla says its model S car will drive itself this summer. *The New York Times*, p. B1. Retrieved from http://www.nytimes.com

Kessler, A. M., & Vlasic, B. (2015, April 3). Semiautonomous driving arrives, feature by feature. *The New York Times*, p. B4. Retrieved from http://www.nytimes.com

Lassa, T. (2013, January). The beginning of the end of driving. *Motor Trend*. Retrieved from http://www.motortrend.com

Lee, D. (2014, August 18). Self-driving lorries "to get UK test in 2015." British Broadcasting Corporation. Retrieved from http://www.bbc.com

Markoff, J. (2014, May 29). Police, pedestrians, and the social ballet of merging: The real challenges for self-driving cars [Bits web log]. *The New York Times*. Retrieved from http://www.nytimes.com

Markoff, J. (2015, April 7). Planes without pilots. *The New York Times*, p. D1. Retrieved from http://www.nytimes.com

Naylor, B. (2013, August 19). Hitting the road without a driver [Web log]. National Public Radio. Retrieved from http://www.npr.org

Neff, J. (2010, May 6). Car accidents claim a life every 15 minutes...and other sobering car crash stats [Web log post]. Retrieved from http://www.autoblog.com

News from Elsewhere. (2014, August 18). Hitch-hiking robot ends 6,000 km journey across Canada [Web log]. British Broadcasting Corporation. Retrieved from http://www.bbc.com

Noe, A. (2015, January 23). The ethics of the "singularity." National Public Radio. Retrieved from http://www.npr.org

Saner, E. (2013, March 30). Will chimps soon have human rights? *The Guardian*. Retrieved from http://www.theguardian.com

Shapiro, A. (2014, June 11). Across Europe, anti-Uber protests clog city streets. National Public Radio. Retrieved from http://www.npr.org

Simon, S., & Bostrom, N. (2015, January 17). Experts petition to keep computers on humanity's side [Interview]. National Public Radio. Retrieved from http://www.npr.org

Tufekci, Z. (2015, April 19). The machines are coming. *The New York Times*, p. SR4. Retrieved from http://www.nytimes.com

US Department of Transportation. (n.d.). Distraction.gov [Website].

Wakefield, J. (2015, March 18). TED 2015: Google boss wants self-drive cars 'for son'. British Broadcasting Corporation. Retrieved from http://www.bbc.com

Wall, M. (2014, October 9). Could a big data-crunching machine be your boss one day? British Broadcasting Corporation. Retrieved from http://www.bbc.com

Ward, M. (2014, September 1). Hi-tech cars are security risk, warn researchers. British Broadcasting Corporation. Retrieved from http://www.bbc.com

Wood, M. (2015, March 20). Video feature: Inside the F015, Mercedes's self-driving car. *The New York Times*, p. B4.

Zakaria, F. (2015). *In defense of a liberal education*. New York, NY: Norton.

7
The Internet of Things

Holiday season marketing (Elliott, 2014; Manjoo, 2014), as well as trendsetting product showcases like the International Consumer Electronics Show (Palmer, 2015), demonstrate the escalating trend of gadgets that talk to each other, learn from each other, even control each other (Rose, 2014).

Most of the focus has been on "smart homes," with furniture, décor, appliances – even toys (Manjoo, 2014) – that direct their own operations and become "housekeepers" (Madrigal, 2015) and perhaps "child care" (Singer, 2015) independent of human intervention. Robots vacuum and mow the lawn (Calamur, 2015). Coffee is ready when you wake, garage doors open just as you arrive, and climate systems maintain the perfect temperature (Wood, 2014). Umbrellas predict storms; trash cans know when a household is running low on products (Rose, 2014). Kitchen appliances write the shopping list, shop, track freshness, measure nutritional values, make food using a 3D printer, find recipes, cook, and clean up (Amos, 2015; Wall, 2013). Forks alert us we eat too fast (Wall, 2013) or chopsticks warn us food is unsafe (BBC Technology Staff, 2014b).

But some system designers think on the larger scale of "smart cities." Several governments and utilities in the US and UK promote "smart meter" programs to conserve resources (BBC Technology Staff, 2015b; Myrow, 2007). After the 2011 earthquake, Christchurch, New Zealand, was rebuilt with embedded sensors to track water quality and leaks, traffic, streetlight functioning, and parking space usage (Murray, 2015). Other innovators aim for the micro-market – implanting technology within human bodies, for example. Some employers are experimenting with implanting identification chips the size of a rice grain under their employees' skin to replace ID card swiping to enter buildings or use other company resources (Cellan-Jones, 2015). Bioelectronics aim to remotely control nerves or help our bodies heal themselves, or to implant Wi-Fi-enabled defibrillators, insulin pumps, or other medical devices (Behar, 2014).

All we need, we are told, is a starter kit to connect everything (BBC Technology Staff, 2015a). But these gadgets may require upgraded wiring inside and outside the home (Madrigal, 2015), and they don't yet smoothly interconnect (Palmer, 2015) as standards have not yet been set (Yu, 2013) even though many companies are vying to be the standard-setter (Wood, 2014). More worrisome for most people is that security is weak. Dangerous hacks and hijacks, for example, of medical devices could occur (Behar, 2014), or entire homes could become ThingBots or spammers via game consoles, refrigerators, and home routers (Belton, 2015; Hu, 2014).

Invasions of privacy arise as televisions, security cameras, and light bulbs spy on us (BBC Technology Staff, 2014a; Wakefield, 2014). Malfunctions in one part of the networked system could cascade into big headaches (Coburn, 2014). Dolls may record children's talk, manipulate their understandings, and alter their notions of friendship (Singer, 2015). Our homes, workplaces, and communities could become 24-hour monitored prisons, or a jumble of expensive technology that can't be integrated – and we might become so dependent on the technology that we forget how to cook, clean, shower, entertain ourselves, or converse with each other. Then, what do we do if there's a blackout?

Daily Life, Automated
with Xiaoyi Cui

A man's voice starts, "Let's celebrate! One hundred years ago, in 1999, Kevin Ashton was the first one to propose the concept of the Internet of Things (IoT) – our current system of every object in the house connected to each other and the Internet to take care of us. Thanks, Genius Ashton!" (Camarinha-Matos, Goes, & Martins, 2013).

It's 7:46 a.m. The radio alarm awakens Sarah. With its sleeping monitor that tracks Sarah's hours and quality of sleep, health indicators, and seasonal patterns, it knows the best time for her to get up. With Sarah still stretching in bed, the curtains draw back to let natural daylight stream in. Sarah listens to the daily news.

As she walks to the bathroom, bedroom speakers go quiet and bathroom speakers get louder. The announcer continues, "IoT turned all our appliances, tools, toys, and utilities into our trusty servants. Can you believe that people used to have to take care of themselves and each other?" (Holler, Tsiatsis, Mulligan, Avesand, et al., 2014). Sarah takes her shower, always the perfect temperature and length. Her bed makes itself. She walks under the automated dryer of warm air that blows from the ceiling. Her parents used to tell her stories of towels – how silly and wasteful to use cloth instead of air!

The kitchen has already prepared and placed coffee, toast, and eggs on the table. The appliances work so well together, Sarah often considers them friends. She sits down at the table, and the newspaper comes up on the table, which doubles as a screen. Every step is in its proper order. Sarah likes reliability. But then, she's never known anything else.

Except for one time. At the last minute, she invited friends over for the evening. It was something her house didn't anticipate. So dinner was served at the small table instead of the larger one for company. Not having her house well trained resulted in some disapproving glances from her friends. It was embarrassing, actually. So Sarah stopped thinking about it.

Sarah switched to a different news feed. It continued reporting the history of the Internet of Things. "The IoT simplified everyone's life. It started small, with health and fitness wearable technology to keep us in good shape" (Acquity Group, 2014). "But the IoT grew rapidly as we, our communities, and our governments wanted smarter bodies, smarter cities, even smarter forests that could detect potential fires (Libelium, 2014). Smarter is better! See how far we've come today! Everywhere you look are sensors and cameras to check the status of everything in real time. We don't have to think about anything."

Sarah's toast doesn't taste as good as expected. "What is going on?" she complains. She checks her refrigerator's meal maker. The nutrition app determined that the mineral balance in Sarah's body was not optimum, so the kitchen salted her toast. The house knows better than Sarah does what is good for her. She sighs, but what can she do? She never learned how to cook. No one knows how to cook anymore.

It's time to prepare for her day. Her closet, which tracks both her calendar and the weather report, gives her the elegant white dress her mother likes, since she will visit her parents this evening. The dress is polyester, which won't wrinkle during the flight to her parents' home. And the closet adds a rain coat: "Eighty percent chance of rain after 2 p.m. after you land." Sarah does not give the information another thought; it is the way it is. She lets the closet remove her robe and slip on the dress. It's time to leave. No need to lock the door or remember keys. There are no keys – something her parents once told her about that she thought would be so annoying to keep track of.

Sarah's driverless car waits at the curb. There is no conversation. The car knows she has a flight in two hours. On the way, a driverless taxi miscalculated the extra braking time needed for a wet road. As her car swerved left then right again, she felt jostled. Not like she could do anything about it, though. She had no control. "Must've not gotten the software update this morning," she thought to herself about the taxi. She had heard some would be taken out of commission because their hardware couldn't keep up with the new rules of the road. "Must be one of those old clunkers," she thought.

The only people she sees at the airport are other passengers. There are no staff. Just one pathway where different sensors read her flight information from the ID chip implanted in her jaw, screen and weigh her luggage, direct her to the proper gate, and welcome her to the flight. It all runs so efficiently, she didn't have time to check her Attendant, the interface with her home that tracks all events in her life. The Attendant said, "You have not seen your parents for six months. You need to ask your father how his cancer treatment is going, congratulate them on their upcoming trip celebrating their 30th anniversary, and report that you are very healthy." Of course, Sarah can check her father's chemotherapy progress and their travel plans online. The world is so transparent now! "It's nice to be reminded and not have to hold all that in my mind," Sarah smiled to herself. "Could you imagine having to carry all that info in my mind? It would give me a headache." She was looking forward to her mother's travel adventure stories. The social media reports didn't quite capture all of her mother's excitement and love of life.

During the flight, her Attendant alerted her to the breaking news that was sent to everyone: "Displaced workers held their third demonstration this year due to high unemployment. The government suggests avoiding these GPS coordinates to not be affected by the unrest." Sarah felt for these unfortunate people. But she didn't understand the agitation. The Internet of Things takes care of everything. Why would anyone want to *work*?

A friend had tried to explain it to her: "Their lives were derailed by innovation and they can't find a place in the new social order. The Internet of Things can do the same work as these people, only faster and better. They feel devalued. They don't qualify for re-education programming, so they can't make a contribution to society any more. They have no reason to be...no purpose."

"I guess that's true," Sarah thought. She was usually pretty agreeable. There was no reason to think critically, or beyond common knowledge.

No one she knew did that. Why make the effort? She had read about how farmers, manufacturers, doctors, lawyers, professors, most jobs were lost once the Internet of Things became fully functional. 3D printers made food. Computer simulations trained people. Search engines and auto-writers presented legal cases. And probes, sensors, and pattern recognition programs diagnosed and treated disease.

The only people who worked were hardware mechanics and software programmers – and not very often, since artificial intelligence took care of most systems' upkeep and upgrades. In fact, the only time she hears about people working in the news is when the elite Cyberguard are deployed due to a hack attack. But even those events are rare since neural sensors are now so sensitive that renegade code could be averted before the idea ever fully crystallized inside the programmer's mind.

"Oh, yeah," Sarah also remembered, "there are also the touchers, the lower class workers who tend the sick or infirm or mentally unstable or criminal individuals... oh, and the artists, who put in effort just for fun." That last thought made Sarah smile. She was an artist. The Internet of Things hadn't yet figured out a way to do such "hands-on" tasks better than humans could.

The news report continues, "Government software reminds the demonstrators that basic needs are available for everyone. There is no need for such emotional outbursts." Behind the announcer, crowds chant, "Men over machines! Rage against the robots!" But the announcer drowns their voices: "Embedded sensors carefully monitor developments. Fencing can be engaged if violence breaks out, to contain the area for safety of others. Repeat: for safety reasons, please do not go to...." Sarah turns down the volume. Her seat senses her sleepiness and leans back automatically.

When Sarah arrives at her parents' house, before the door automatically opens, the auto-butler warns her, "Do not quarrel with Mr. James. Avoid talk of romantic relationships and career. Thank you." Her parents are on the veranda, following the fitness software's yoga poses. Her mother is having trouble with the pose, which the software senses and adjusts to a less strenuous pose. She can tell that her parents have been a little troubled by something because they set their house décor to reflect their mood. Last time she was here, the walls were orange, and now they were a more subdued ivory.

"Sarah!" Her father saw her first. They exchanged pleasantries. She accidentally mentioned her recent studio piece. Her father scowled, "Your studio? I told you, you should be a software developer. Did you see the demonstration today? Why do you want to be an artist? Machines can draw. Nobody draws by themselves anymore."

"Children do," Sarah responded. "They love it. They think making stuff is unique and great! So do older people that remember the days before machines did everything. They get restless with all this 'free' time."

Her mother changed the subject: "Have you met someone?" Sarah shook her head, "The dating algorithm has not found anyone. Nobody matches me; I don't want to waste my time. I don't need a partner. My house suits me fine. It knows my personality and my needs very well. The voice that wakes me up is sexy. And, he is not a troublemaker."

After dinner, Sarah practices Chinese brush painting, does her stretching exercises, then yawns. The sensor notes her sleepiness, turns on the evening hygiene regimen, warms the bed, closes the curtains, and dims the lights. The next morning, her routine begins again. Sarah likes regularity. The only startle the next day was, as she arrived back at her own house, her Attendant notified her about a message from her airline: "We apologize about today's inconvenience. Flight RQ905's system was hacked. My pleasure to announce that we found and solved the problem immediately and nobody got hurt. Thank you for traveling with us. Looking forward to serving you again in the future."

Over dinner, she reviewed the day's news. One headline stood out: "ART OUTLAWED." Sarah felt confused. She read: "To reduce the wasting of resources for frivolous uses, all human art-making is prohibited. Printers can provide precision copies of masterworks, and randomization software can devise 'novel' combinations of color and pattern efficiently. The waste of people's time to paint the same vision in different ways to 'get it right' is ineffective. Art-making will be deleted from all records and will be forgotten."

Further exploration

1. Who might not prosper if gadgets do everything for people? How might they survive?
2. How could the Internet of Things affect privacy? What might be the ripple effects of those interactive effects?
3. As gadgets take over more tasks, what should people do in their new free time?

References

Acquity Group. (2014). The Internet of Things: The future of consumer adoption. Retrieved from http://www.acquitygroup.com
Amos, J. (2015, April 14). "Robot chef" aimed at home kitchen. British Broadcasting Corporation. Retrieved from http://www.bbc.com
Behar, M. (2014, May 25). Invasion of the body hackers. *The New York Times*, p. MM36.
BBC Technology Staff. (2014a, July 30). Smart home kit proves easy to hack, says HP study. British Broadcasting Corporation. Retrieved from http://www.bbc.com
BBC Technology Staff. (2014b, September 4). "Smart chopsticks" unveiled in China. British Broadcasting Corporation. Retrieved from http://www.bbc.com
BBC Technology Staff. (2015a, February 24). Internet of Things starter kit unveiled by ARM and IBM. British Broadcasting Corporation. Retrieved from http://www.bbc.com
BBC Technology Staff. (2015b, March 27). Smart meter scheme could be IT disaster, says IoD. British Broadcasting Corporation. Retrieved from http://www.bbc.com
Belton, P. (2015, February 17). Is your toaster a silent recruit in a "thingbot" army? British Broadcasting Corporation. Retrieved from http://www.bbc.com
Calamur, K. (2015, April 16). We might welcome robot lawn mowers, but astronomers aren't so happy [The Two-Way web log post]. National Public Radio. Retrieved from http://www.npr.org
Camarinha-Matos, L. M., Goes, J., & Martins, J. (2013). Contributing to the Internet of Things. *Technological Innovation for the Internet of Things* (IFIP AICT Series 394), pp. 3–12. Retrieved from http://www.academia.edu
Cellan-Jones, R. (2015, January 29). Office puts chips under staff's skin. British Broadcasting Corporation. Retrieved from http://www.bbc.com
Coburn, T. (2014, June 14). Our dream of a connected home could become a nightmare. *Wired*. Retrieved from www.wired.com
Elliott, S. (2014, October 23). A tech twist on home for the holidays. *The New York Times*, p. B4. Retrieved from http://www.nytimes.com
Holler, J., Tsiatsis, V., Mulligan, C., Avesand, S., Karnouskos, S., & Boyle, D. (2014). *From machine-to-machine to the Internet of Things: Introduction to a new age of intelligence*. Oxford, UK: Academic Press.
Hu, E. (2014, January 16). What do you do if your refrigerator begins sending malicious emails? [Web log post]. National Public Radio. Retrieved from http://www.npr.org
Libelium. (2014). 50 sensor applications for a smarter world. Retrieved from http://www.libelium.com
Madrigal, A. (2015, February 25). The world loves the smartphone. So how about a smart home? [Web log post] National Public Radio. Retrieved from http://www.npr.org
Manjoo, F. (2014, November 27). Tech toys that go beyond the screen. *The New York Times*, p. B1. Retrieved from http://www.nytimes.com
Murray, S. (2015, January 28). The humble lamppost helps to shine a light on smart cities. *The Financial Times*, p. 2.

Myrow, R. (2007, January 22). California hopes smart meter will spur conservation. National Public Radio. Retrieved from http://www.npr.org

Palmer, M. (2015, January 28). Expect the spectacular – but just not yet. *The Financial Times*, p. 1, 2.

Rose, D. (2014). *Enchanted objects: Design, human desire and the Internet of Things*. New York: Scribner.

Singer, N. (2015, March 29). A Wi-Fi Barbie doll with the soul of Siri. *The New York Times*, p. BU3. Retrieved from http://www.nytimes.com

Wakefield, J. (2014, July 8). Smart LED light bulbs leak Wi-Fi passwords. British Broadcasting Corporation. Retrieved from http://www.bbc.com

Wall, M. (2013, November 25). Food bytes: The kitchen goes digital. British Broadcasting Corporation. Retrieved from http://www.bbc.com

Wood, M. (2014, June 12). Controlling your smart home with one hub. *The New York Times*, p. B8.

Yu, A. (2013, December 13). Tech companies take step toward the "Internet of Things." National Public Radio. Retrieved from http://www.npr.org

8
Drones

Drones are taking off in civilian life. Some of the clever uses include restaurant services (News from Elsewhere, 2014; Slane, 2014; Wong, 2015), documenting ephemeral beach art (Murphy, 2014a), monitoring mountain yaks (News from Elsewhere, 2013), contraband delivery into prisons (Schmidt, 2015), and Internet access in remote areas of the world (Hardy & Goel, 2015). Industrial and individual interest in drones has skyrocketed to the extent that governments around the world are pressured to regulate them (BBC Technology Staff, 2015; De la Baume, 2015; Neuman, 2015; NPR Staff, 2014; Peralta, 2014).

The FAA first approved the use of drones for profit in the US for energy exploration (Chappell, 2014b). Rules issued in 2015 restricted commercial drones to less than 55 pounds, and their use to daylight hours, always within eyesight of a licensed operator, slower than 100 mph or lower than 500 feet, and not near airports, other aircraft, or populated areas (Neuman, 2015). Filmmakers and retailers recently received exemptions (Barnes, 2014; Wingfield, 2015), opening the gates for wider use, such as photojournalism (Hu, 2014; Shahani, 2014), product delivery (Scott, 2014; Streitfeld, 2014), and disease tracking (Akpan, 2014).

All these uses – plus drone hobbyists' personal use – take intrusion by government, companies, and even individuals to new heights. Not only are humans' personal spaces invaded, but animals are being crowded out of airspace (Morelle, 2015). As lawmakers and regulators struggle to keep up, ethics becomes paramount. How could these all-seeing robots buzzing around us change our personal safety and social responsibility? How can we make sure they are put to GOOD use?

Insights from Above
with Victoria Westerband

Unmanned aerial vehicles, better known as drones, are remote-controlled machines that may have cameras, sensors, facial recognition, cargo bays, recorders, speakers, or other features (Rose, 2014). They offer lower cost as well as less detectability than human investigators. But these advantages are also drawbacks, making drone use easier to cross ethical lines of fairness and avoidance of harm.

First used in the military to attack the enemy without risking the lives of our own soldiers, drones have been adapted to use in policing and security (Kelion, 2014), search and rescue (BBC Technology Staff, 2014b; Rose, 2014), agriculture (Lowe, 2014; Runyon, 2015), forestry and firefighting (Palmer, 2014), wildlife monitoring (News from Elsewhere, 2013; Wall, 2014), photography (Hu, 2014; Murphy, 2014a), and archaeology and geography (Ames, 2015; Neuman & Blumenthal, 2014). Most have been for official uses or limited to out-of-the-way locations. But now, drones are entering more populated areas, like over Paris (e.g., De la Baume, 2015), and more types of businesses want to use them.

Furthermore, a 2012 law allows wider use of drones in US airspace (Wingfield & Sengupta, 2012), and individuals now have easy access to buy personal drones (Wingfield, 2014). Drones offer more capacities than older hobbyist model planes (Murphy, 2014b; Wingfield, 2014), but still, for the most part, are governed by older rules (FAA, 1981). Building, owning or flying drones recreationally is legal during daylight, at low altitudes and pilot controlled at all times, and away from other people and airports (FAA, 1981). But proliferation of drones could lead to personal injuries (BBC Technology Staff, 2014a; Rose, 2014), privacy concerns (Sydell, 2015), and security breaches (Kelion, 2013; Schmidt & Shear, 2015).

Personal safety

On the brighter side, drones could help prevent injury or, if injury occurs, provide faster care or information delivery to authorities. For example, home drones could help watch children, perhaps preventing drowning accidents, sibling fights, or break-ins. If crashes, terrorist attacks, or contagious disease outbreaks occur, drones could be first on the scene, which is often dangerous for humans, to gather information and relay it to police or doctors.

Still, most people worry about how drones can inflict harm. For example, high-speed pursuits could strike people who happen to be in the way. Drones nearly struck someone walking in Manhattan and an athlete in Australia (BBC Technology Staff, 2014a; Hoffer, 2013). Pilots can lose control and crash drones (Shear & Schmidt, 2015). If used near roads, they could distract drivers, contributing to accidents. The FAA has already heard reports of drones near airplanes that distract pilots (Wingfield, 2014).

Privacy

Current US law does not protect personal privacy if we are out in public or visible from a public space (Wingfield & Sengupta, 2012). Small drones can get into remote areas or tight corners, dangle outside windows, and hover over our gardens enclosed by 10-foot walls. The law also does not prohibit drones from flying over our homes (Kleinman, 2014; Sydell, 2015). Drones can follow us. The American Civil Liberties Union claims that drones breach civil rights by monitoring us even walking down a street (Khaki, 2012). Thus, drones have the potential to surreptitiously make a space, which without drones most people considered private space, visible from a public space. Paparazzi have stalked celebrities for decades. With drones, we all potentially become fair game.

Our electronic gadgets can betray our privacy as well. Drones could be programmed to steal data from smartphones (Gittleson, 2014). The method is not new, but drones make the use of the hack more mobile. It is also possible that drones can create new avenues for hackers to infiltrate cars and homes that have connected to the Internet. Current drone programming most likely does not include an ethical code. The software does not specify moral boundaries, moral norms, or ethical criteria for selecting appropriate drone behavior in situations that pose moral dilemmas (a similar discussion has arisen in relation to driverless cars).

The White House called for a regulatory framework to address privacy, accountability, and transparency in drone-collected data (Office of the Press Secretary, 2015). That process will take time. In the interim, social entrepreneurs have started services to provide conscientious drone pilots with a list of locations not to fly over that they can program into their drones – similar to the Do Not Call list for telemarketers (Sydell, 2015). A designer devised an "anti-drone hoodie" to deflect drones' transmission or collection of digital data (Meltzer, 2013). Other enterprising activists may be launching similar services or products to make drones more "aware" of privacy issues.

Security

Drones can affect two kinds of security: digital security and event security. As with computer operating systems, the more widely used drones are, the more popular they become to hackers because the bigger payoff drones can make if their infiltration is successful. If drones become as ubiquitous as computers, mobile devices, credit cards, and other repositories of valuable personal information, we might expect drone "skyjacks" to increase (Kelion, 2013).

Privacy infringement betrays our secrets, whereas security breaches betray how gadgets operate. The pilot of the drone that hit the Australian athlete asserted that his drone had been hacked (BBC Technology Staff, 2014a). A security researcher demonstrated how he hijacked a popular brand of drone (Kelion, 2013). Encryption, authentication, secret keys, and other security tools need to be implemented to ward off unauthorized control of a drone.

Security also concerns personal security in public spaces. Drones spotted over Paris put the whole city on alert (De la Baume, 2015). Law enforcement agencies have added "vertical security" issues to their planning for big events now that they must consider drones in the sky. A drone that hovered a flag over a European soccer match led to a riot (Wingfield, 2014). As a result, the FAA updated its policy so that drones could not fly near sporting events. Although these instances are localized, the concern is that, even though laws require drone pilots to be within sight of the drone, drone technology is not limited to such short distances. The perpetrator of an incident could be long gone by the time damage ensues.

Prosocial perspectives

Drones are small, light-weight, and agile, so they can promote mobility that can bring us closer and speed up positive change (Thackara, 2005). If drone pilots can move beyond selfish motives to *get* information or profit or an advantage of some type, then a variety of opportunities could be considered to use drones to *give* prosocial benefits, such as health or companionship.

On a global scale, drones might be put to good use for lifesaving missions. The United Arab Emirates sponsors an international competition for the "Drones for Good Award" (United Arab Emirates Prime Ministers Office, 2015). The website describes inspiring ways to use drones to improve individual, community, society, and global wellbeing. Ideas have included manufacturing hardier drones that can resist collision damage and waterproof drones that could be lifeguards in dangerous waters. Drones could be used to extend or amplify cellular coverage during emergencies, deliver medicines or medical devices, detect landmines, and counteract pollution (Chhabra, 2015; Eng, 2013).

Drones might also be considered aesthetically. Although few people would consider current drones beautiful, it is possible that later drone designs include aesthetic as well as functional criteria. Furthermore, the assembly and interaction of drones could delight our aesthetic sensibilities. For example, researchers created drones that self-direct their own movements into coordinated formations (Chappell, 2014a), which might herald an artistic form of robot aerobatics, circus, or ballet.

On a personal scale, drones could become friendlier. They might serve as nurse assistants, especially in contagious zones, to provide not only health status monitoring, but also music, warmth, pleasant scents, and other forms of comfort to sick individuals in quarantine. Although they should not replace parental care or supervision, drones might be short-term nannies to accompany children walking home after school, crossing streets, and avoiding trouble. Some personal drones could be considered "pets" for families or for elderly, disabled, or homebound individuals. They offer protection, assistance and company, just as a dog might. Pet drones could alleviate loneliness and assist in activity or medical monitoring.

Of course, these prosocial uses of drones must carefully consider the safety, privacy, and security issues explored above. Since these prosocial drones become trusted household members, hacks into their systems could be particularly damaging. In these instances, hacked drones could be felt as betrayals with long-term repercussions to future social interactions as people do not easily rebound from breaches of trust. For example, drone pets are likely to be accepted and taken for granted within the home. The information hackers intercepted could be much more personal than credit card numbers.

Broader horizons

The appeal of drones is that they provide new and perhaps better vantage points for observing and understanding our world. They promise to provide more accurate data, pictures, and mapping, especially for locations that are hard for humans to reach or situations that are dangerous for humans. Such conditions include remote areas with poor infrastructure, war zones, contagious areas, and undisturbed wilderness, among other possibilities.

From the perspective of the drone pilot, this added perceptual capability could be exhilarating, curiosity-satisfying, perhaps even an extension of the self. The pilot experiences a viewpoint previously not possible, except perhaps through hot-air ballooning or bungee jumping. With drones, the pilot has *control*. However, from the perspective of the observed – at least for humans who are aware of being gazed upon – drones introduce *loss of control*. Drones are small, nimble, and can fly almost undetected. Thus, they can breed anxiety and uncertainty: Are we being watched, and if so, by whom? Drones don't announce who or where the pilot is. And why? Drones don't disclose their purposes.

Thus, the broader perception that drones provide creates a tension between seeing and being seen. Individuals and organizations need to be wary of, and perhaps protect themselves from, others' surveillance while simultaneously considering how their own drone use can enhance their knowledge of the world. Much of the regulatory discussion aims to balance these perspectives. But regulation is not foolproof. Drones may be a good example of why and how ethics are important – how individuals, within our own minds, recognize and appreciate that multiple perspectives are in play, and that we need to be ethically conscientious that we are part of a larger society in which others also have rights, needs, and liberties.

Further exploration

1. Who is responsible for drone actions and effects? Rank those involved in terms of responsibility, and consider what types of responsibility they have.
2. How could drone use affect the pilot's sense of identity and power?
3. What mechanisms, besides regulation, could promote the ethical use of drones?

References

Akpan, N. (2014, October 22). Drones are taking pictures that could demystify a malaria surge. National Public Radio. Retrieved from http://www.npr.org
Ames, J. (2015, February 14). Brazil Amazon: Drone to scan for ancient Amazonia. British Broadcasting Corporation. Retrieved from http://www.bbc.com
Barnes, B. (2014, September 26). Drone exemptions for Hollywood pave the way for widespread use. *The New York Times*, p. B1. Retrieved from http://www.nytimes.com
BBC Technology Staff. (2014a, April 7). Australian triathlete injured after drone crash. British Broadcasting Corporation. Retrieved from http://www.bbc.com
BBC Technology Staff. (2014b, July 25). Drone operator explains how he found missing man. British Broadcasting Corporation. Retrieved from http://www.bbc.com
BBC Technology Staff (2015, January 27). Thailand mulls jail term for unlicensed drone pilots. British Broadcasting Corporation. Retrieved from http://www.bbc.com
Chappell, B. (2014a, February 26). Robot swarm: A flock of drones that fly autonomously. National Public Radio. Retrieved from http://www.npr.org
Chappell, B. (2014b, June 10). Drones approved: FAA gives ok to first commercial use over land. National Public Radio. Retrieved from http://www.npr.org
Chhabra, E. (2015, February 27). Drones for good: Projects from around the world on how drones can help us. *Forbes*. Retrieved from http://www.forbes.com
De la Baume, M. (2015, February 26). French hold 3 journalists in drone episodes, but no link to earlier flights is seen. *The New York Times*, p. A10.
Eng, K. (2013, June 11). Speedy delivery: Andreas Raptopoulos at TED Global 2013 [Web log comment]. Retrieved from http://blog.ted.com
Federal Aviation Administration (FAA) (1981, June 9). Model aircraft operating standards (advisory circular 91–57). Washington, DC: Author. Retrieved from http://www.uavm.com/images/ac91-57.pdf
Gittleson, K. (2014, March 28). Data-stealing Snoopy drone unveiled at Black Hat. British Broadcasting Corporation. Retrieved from http://www.bbc.com
Hardy, Q., & Goel, V. (2015, March 26). Drones beaming web access are in the stars for Facebook. *The New York Times*, p. B1. Retrieved from http://www.nytimes.com
Hoffer, J. (2013, October 3). Small drone crash lands in Manhattan. *ABC Eyewitness News*. Retrieved from http://abclocal.go.com/wabc/story.
Hu, E. (2014, May 5). Drone journalism can't fully take flight until regulators act. National Public Radio. Retrieved from http://www.npr.org
Kelion, L. (2013, December 4). Parrot drones "vulnerable to flying hack attack." British Broadcasting Corporation. Retrieved from http://www.bbc.com
Kelion, L. (2014, June 18). African firm to sell pepper-spray bullet firing drones. British Broadcasting Corporation. Retrieved from http://www.bbc.com
Khaki, A. (2012, February 21). Domestic drones: Big Brother's prying eyes in the sky [Web log comment]. Retrieved from http://www.aclu.org
Kleinman, Z. (2014, November 18). Topless sunbather photographed by estate agent drone. British Broadcasting Corporation. Retrieved from http://www.bbc.com

Lowe, P. (2014, July 19). Deploying drones to get an overview of factory farms. National Public Radio. Retrieved from http://www.npr.org

Meltzer, T. (2013, March 31). The anti-drone hoodie that helps you beat Big Brother's spy in the sky. *The Guardian*. Retrieved from http://www.theguardian.com

Morelle, R. (2015, May 1). "Airspace reserves" could protect wildlife. British Broadcasting Corporation. Retrieved from http://www.bbc.com

Murphy, K. (2014a, July 27). Andres Amador. *The New York Times*, p. SR2.

Murphy, K. (2014b, December 7). Things to consider before buying that drone. *The New York Times*, p. SR5. Retrieved from http://www.nytimes.com

Neuman, S. (2015, February 15). Commercial drone rules to limit their weight, speed and altitude. National Public Radio. Retrieved from http://www.npr.org

Neuman, W. & Blumenthal, R. (2014, August 14). New to the archaeologist's tool kit: The drone. *The New York Times*, p. A10. Retrieved from http://www.nytimes.com

News from Elsewhere. (2013, December 13). China: Drones used to monitor yaks. British Broadcasting Corporation. Retrieved from http://www.bbc.com

News from Elsewhere. (2014, May 23). Police investigate pizza deliveries by drone. British Broadcasting Corporation. Retrieved from http://www.bbc.com

NPR Staff. (2014, May 5). FAA Head: Safety, privacy concerns abound in regulating drones. National Public Radio. Retrieved from http://www.npr.org

Office of the Press Secretary. (2015, February 15). FACT SHEET: Promoting economic competitiveness while safeguarding privacy, civil rights, and civil liberties in domestic use of unmanned aircraft systems. Washington, DC: The White House.

Palmer, K. M. (2014, August). Firefighting firepower airborne help for blazes. *Wired*, p. 42.

Peralta, E. (2014, June 21). National Park Service temporarily bans drones in national parks. National Public Radio. Retrieved from http://www.npr.org

Rose, D. (2014, November). Dudes with drones. *The Atlantic*, pp. 86–90.

Runyon, L. (2015, March 26). Is Colorado primed to become the Silicon Valley of agriculture? National Public Radio [The Salt web log]. Retrieved from http://www.npr.org

Schmidt, M. S. (2015, April 23). Airmail via drones is vexing for prisons. *The New York Times*, p. A13. Retrieved from http://www.nytimes.com

Schmidt, M. S., & Shear, M. D. (2015, January 29). Drones spotted, but not halted, raise concerns. *The New York Times*, p. A1. Retrieved from http://www.nytimes.com

Scott, M. (2014, September 25). DHL to begin deliveries by drone in Germany [Bits web log]. *The New York Times*. Retrieved from http://www.nytimes.com

Shahani, A. (2014, May 16). Are filmmakers using drones illegally? Looks like it. National Public Radio. Retrieved from http://www.npr.org

Shear, M. D., & Schmidt, M. S. (2015, January 28). White House drone crash described as a U.S. worker's drunken lark. *The New York Times*, p. A15. Retrieved from http://www.nytimes.com

Slane, K. (2014, November 19). *Mistletoe drones? TGI Fridays gets into holiday spirit in weirdest way possible*. Boston.com. Retrieved from http://bit.ly/14MWTSQ

Streitfeld, D. (2014, July 12). Amazon asks permission from F.A.A. to test drone delivery system. *The New York Times*, p. B7.

Sydell, L. (2015, February 23). No you can't sign up to keep drones away from your property. National Public Radio. Retrieved from http://www.npr.org

Thackara, J. (2005). *In the bubble: Designing in a complex world.* Cambridge, MA: The MIT Press.

United Arab Emirates Prime Ministers Office. (2015). The UAE Drones for Good Award. Retrieved from http://www.dronesforgood.ae

Wall, M. (2014, July 21). Can drones help tackle Africa's wildlife poaching crisis? British Broadcasting Corporation. Retrieved from http://www.bbc.com

Wingfield, N. (2014, November 27). Now, anyone can buy a drone. Heaven help us. *The New York Times*, p. A1. Retrieved from http://www.nytimes.com

Wingfield, N. (2015, March 20). Amazon wins approval to test delivery drones outdoors. *The New York Times*, p. B2. Retrieved from http://www.nytimes.com

Wingfield, N., & Sengupta, S. (2012, February 18). Drones set sights on U.S. skies. *The New York Times*, p. A1. Retrieved from http://www.nytimes.com

Wong, T. (2015, February 8). Drone waiters to plug Singapore's service staff gap. British Broadcasting Corporation. Retrieved from http://www.bbc.com

Part III
Body Shapers

9
Preventive Gene Testing

To be forewarned is to be forearmed. That is the idea behind preventive gene testing. If we know in advance that our genes predispose us to a specific disease, we can do something about it. Perhaps with screenings and earlier treatments, eventually doctors may be able to directly alter our genes so we – or our children – never develop the disease. Although gene-editing is consider one of the most heinous ethical breaches among scientists today, small-scale experiments still occur (Kolata, 2015).

The proliferation of preventive genetic testing has far outpaced regulation, or even informed discussion. Do-it-yourself kits can be sent directly to consumers (Alsever, 2011), despite controversy whether it is a medical test requiring a doctor's prescription (Pollack, 2013). People without medical training may not understand what the test results mean (Pear, 2008), and even medical researchers haven't figured out how to translate many genetic mutations into real-life implications (Lewin, 2000). So these tests can breed more confusion and anxiety than answers and progress (Grady & Pollack, 2014). Furthermore, who owns the genetic information? Databases of personal genomes are already being assembled (Pinker, 2009). Do we want to know? Who should know? How much? And what do we do with this knowledge of who we are genetically?

Our New Fortune Teller?

with Victoria Russo

Announcer: Up next...*Mad Science* for May 15, 2040...would YOU want to know what diseases you'll get 30 years from now? Stay tuned!

Dailey: Welcome! On tonight's episode of *Mad Science*, a genetics specialist recalls how *preventive* genetic testing, since its introduction back in the 2000s, has changed and continues to change how we live. Welcome, Dr. Michael Santos! Genes are...well, *us*, right? How do genes work, and how do we know how they work?

Santos: Our genes are the blueprints for our bodies. Different configurations produce different features, like hair or skin color, or different organs, like the heart or the liver, or different functions, like digestion or sleep.

At the turn of the century, the large-scale Human Genome Project mapped all human genes (www.genome.gov). After that, a lot of research investigated genetic mutations related to diseases. Not only genetic diseases like Down's syndrome, but also heart disease, cancers, and Alzheimer's disease. Researchers and doctors thought, hey, maybe we could use this knowledge to *prevent* diseases (Evans, Skrzynia, & Burke, 2001). So having personal genomes tested became popular (Lewin, 2000). In particular, these people wanted to know if they would get a scary disease later in life.

Dailey: How is that done?

Santos: It's simple, really. Take someone's blood, hair, skin – even a swab inside the mouth. Send it to a lab, and they send you back a report that tells you whether you have gene mutations that may increase your likelihood of particular diseases (National Institutes of Health, 2006). At first these tests were expensive – a "luxury good" (Agus, 2013; Harmon, 2008) – but costs decreased rapidly (Stein, 2015), especially after companies could no longer patent genes (Grady & Pollack, 2014). Genetic mapping tests were marketed directly to consumers (Alsever, 2011), although not without controversy (Pollack, 2013).

Dailey: So if I have my whole genome sequenced, I will know *every* disease I will get over my lifetime?

Santos: Not quite. At least, not yet (Centers for Disease Control and Prevention [CDC], n.d.). Or maybe never, because, as we know now, genes are your identity, but they are *not* your destiny. There have been fascinating developments in medical care ever since we realized "personalized medicine" was possible, concocting designer drugs tailored to a particular person's genome (Associated Press, 2015; Pear, 2015). Now it's common for doctors to run gene maps on patients, especially for cancer tumors (Harris, 2015). The idea is for pharmacists and chemists to devise medicines just for that patient's body chemistry. Looking back, it was a bit barbaric to think random clinical trials for new pharmaceuticals was the "gold standard." These trials could only understand "average" drug effects or effects on the "average" person. But there's no average person! Everyone has different chemistry. Average can be far from optimum for any specific individual.

Dailey: There were so many reports of early patients who suffered through treatments that didn't work in those early days. People were misdiagnosed or diagnosed too late. Looking back, it seems like doctors were just guessing half the time. Back before we understood what "identity" actually meant. But also, people would receive their genome maps and not have a clue what they meant (Kolata, 2012; Pear, 2008). Could you talk about that?

Santos: Yes, at that time, the test results were just lists of genetic mutations – was I a carrier or not? Even the experts weren't sure what the mutations led to (Kolata, 2012), except for a few diseases like Huntington's, where a carrier had 100% chance of getting the disease (Evans et al., 2001). There was a lot of interest in breast and ovarian cancer, with the discovery of specific genes (Jolie, 2013; Rabin, 2013). But, for most mutations, back then, knowing your genome provided little guidance on what you should do next. Just screening more often created more anxiety than relief (Grady & Pollack, 2014; Lerman, Croyle, Tercyak, & Hamann, 2002).

There was a lot of optimism and speculation (Pinker, 2009) and calls for caution (Harris, 2015; Pear, 2015). It took researchers several years to work out the intricate dance between a particular person's body composition and biochemistry, on one hand, and gene, stem cell, nuclear, implant, and microbiome therapies, on the other hand. And we may never be "done" with that process because we continue to evolve (American College of Preventive Medicine [ACPM], n.d.; Brody, 2010).

Dailey: Of course, preventive gene testing at birth is required by law for everyone now. So let's move beyond the medical developments into social developments related to preventive gene testing. For example, I read, half a century ago, that knowing your genome could cause you problems with bias and discrimination (Harmon, 2007).

Santos: Well, the belief was that since we can look at a map of anyone's genes, perhaps employers or insurers could predict who would be "less risky" (Blakeslee, 1990). There was also talk about use of genes for job matching – whether their employees' genes are a good fit for the line of work they are in. People's genetic maps that reveal certain qualities may be more apt to get top-of-the-line work.

But everyone's genetic maps are all public now, as predicted it would become (Lerman et al., 2002). Everyone is in one database – an offshoot of a research project started back in the 2000s (Pinker, 2009) – and genetic discrimination became illegal in 2008 (ACPM, n.d.).

Dailey: I think it's much better that everyone is involved. When it was an individual decision, someone on their own, and the results showed a disease-causing gene mutation back then, it was hard.

Santos: Yes. Should I tell my family? Should I have children? Donate sperm? (Lewin, 2000). Agonizing choices. And do I really want to know? (Kolata, 2012). There is rarely one-to-one correspondence between a particular gene and a particular disease. Life just doesn't seem to work that way. The body is dynamic.

Dailey: But it didn't take long for genetic testing to become a fundamental tenet of our society to define who we are, right?

Santos: An interesting development is how these databases of personal genomes became popular because they were integral to hobbyist genealogy – to find out who you are related to – past and present (Harmon, 2006; Johnson, 2015). Although some hobbyists didn't like the change because it took the fun detective work out of building a family tree, it seems like aggregating genome testing has done wonders to help us realize that we really are – eventually – one big human family!

Dailey: But it also helped some people, for example, who discovered they carried genes of a particular race or ethnicity that was eligible for aid or benefits (Harmon, 2006).

Santos: Or, back to medicine, for treatments that showed particularly good results for people of particular ancestry (Harmon, 2006). One change in perspective that I find particularly hopeful, which came from tracking specific families, is that researchers realized not only was a *treatment* mentality outdated, but so too was a *preventive* mentality. Because we realized that mutations aren't "bad." Living cells have been mutating from the beginning. Life is change, adaptation, opportunism. The new insight was that mutations can be protective against diseases or aging (Kolata, 2014). So many diseases that used to be big worries – HIV, heart disease, diabetes, early onset Alzheimer's – we're beginning to use protective genetic mutations preemptively.

Dailey: As you've said, it's been about 50 years since this medical innovation launched. Switching from looking back to looking forward, what are the promises and worries?

Santos: Most people think of the worries first. What could go wrong? A couple of unsettling developments I see...First, preventive genetic testing may be keeping us from developing easier methods to promote lifelong health. Too much emphasis on only genes negates the roles of environment and even a person's own behavior. Several studies suggest that alarming test results shock us at first, but don't change most people's habits to be healthier (Heshka, Palleschi, Howley, Wilson, & Wells, 2008). Second is overreliance on genes to identify criminals.

Dailey: But that saved governments a lot of money. No more police randomly roaming the streets to, hopefully, intercept "bad guys" before they struck. It is much more efficient to use genetic testing – police, as you know, are now primarily lab technicians – to prevent crime by knowing the genetic proclivities to engage in illegal activities.

Santos: But the process is not perfect and far from fair in some cases because someone's gene – considered "faulty" at the time – condemned the person to death. If a person had a bad gene, there was no hope of a good future. That belief continues. So they are put to death preemptively.

Dailey: But I see a lot of other positive possibilities. Now we don't identify a disease and battle it – the language was always war imagery. We tried to cut short the disease trajectories by removing the genetic lines of sick families. Even when it seemed to work, and our prevention methods eradicated a disease from the population, it came back, maybe in a different form because the new versions were more like "cousins" or "hybrids."

Santos: When we thought in terms of war, we placed ourselves at war with...ourselves. Our newer thinking is to embrace life's processes – including what used to be called "disease" because it didn't "feel well." Now we understand that "it" is "us." Disease is not foreign. Even with an infection, it is our cells reacting to the invader that often cause symptoms. The cells are me. They can change. I change. Yes, there can be suffering, but we have developed good pain protocols for that. A disease is a problematic interaction within or between life forms. But that's how life – in the big picture – learns. Some diseases are incurable still. We are all part of that dance.

Dailey: Back to the intricate dance you mentioned before.

Santos: Yes. A dance of genes, nutrients, opportunities...

Dailey: But then, why have treatments at all?

Santos: To be the best me. To grow, move forward. Keep up life's momentum. Especially for self-aware life forms like us. It's important to keep going, experience life. The assumption was that genes were deterministic. They told us the end of the story. But they are only the launch pad for a human life.

Dailey: Doesn't that insinuate "designer" people? Maybe even "designer" babies? It's not a far leap to want to control others becoming the best them from my own point-of-view. Like the standard "kiss date" ritual. After that first kiss, potential lovers immediately go home to take a genetic test of the other person's DNA left on the lips (DeVito, Shamberg, Sher, & Lyon, 1997). To glimpse an insight into their personalities, see what kinds of children they'd have, how long their partner would live. Genes as a crystal ball!

Santos: Well, our education system *is* organized by these genetic maps, and there is some talk of implementing marriage laws based on genetic maps of parents to foresee what their children will be like. Of course, way back in the 20th century, it was a crude measure, but people had to get blood tests to get married.

Dailey: So we pick our mates based on how we want our children to turn out as opposed to picking our mates because we love them?

Santos: What is "love"? We've discovered it's a hormone – oxytocin. So why not set my sights on someone who is genetically predisposed to produce more oxytocin – to love me more, or to have more loving, caring, prosocial children?

Dailey: But don't you feel like this exploits the child – whether it is to be a genius child or a moral leader? Don't you think this would have consequences on how the family is structured and how a family operates? Isn't there a chance that children will be aware that they were created not as their own authentic selves but for some instrumental reason (Ishiguro, 2005)? What if society goes through "genetic fads" where different genotypes or phenotypes come in and out of fashion? Someone is stuck with their genome for life – at least for now. It would be an awful experience to have an "out of fashion" genome!

Santos: While there is the risk of continuing harmful social trends that have been around for a long time – stereotyping, prejudice, ostracism – it is also possible to use preventive gene testing to maintain and promote human diversity. Some families and ethnicities at risk for extinction due to genetic diseases could be saved, their ways of life preserved. What's more, preventive testing could help us see our best opportunities, not just what to avoid but what to seek. Just as our genes can put us "at risk" for disease, we also can be "at risk" for growth, success, joy.

Dailey: Well, thank you, Dr. Santos. This conversation has been quite enlightening. It's time now for our listeners to call in. What do *you* think are the ethical implications of preventive gene testing into the future?

Further exploration

1. What are the harms, costs, or conflicts of trying to prevent something we fear might happen from actually happening?
2. How much do genes make up who we are? What is the relationship of genes to identity? How might preventive gene testing keep us from being or becoming who we are?

References

Agus, D. B. (2013, May 20). The outrageous cost of a gene test. *The New York Times*. Retrieved from http://www.nytimes.com

Alsever, J. (2011, November 3). Do-it-yourself DNA testing. *Prevention*. Retrieved from http://www.prevention.com

American College of Preventive Medicine [ACPM]. (n.d.). Genetic testing clinical reference for clinicians. Washington, DC: Author. Retrieved April 21, 2015 from http://www.acpm.org/?GeneticTestgClinRef

Associated Press. (2015, January 30). Obama calls on Congress to fund "precision medicine" studies. *The New York Times*. Retrieved from http://www.nytimes.com

Blakeslee, S. (1990, December 27). Ethicists see omens of an era of genetic bias. *The New York Times*. Retrieved from http://www.nytimes.com

Brody, J. E. (2010, April 12). Cancer survival demands steady progress. *The New York Times*. Retrieved from http://www.nytimes.com

Centers for Disease Control and Prevention [CDC]. (n.d.). Genomic testing. Retrieved from http://www.cdc.gov

DeVito, D., Shamberg, M., Sher, S., & Lyon, G. (Producers) & Niccol, A. (Director). (1997). *Gattaca* [Motion picture]. United States: Columbia Pictures.

Evans, J. P., Skrzynia, C., & Burke, W. (2001). The complexities of predictive genetic testing. *British Medical Journal, 322*(7293), 1052–1056.

Grady, D., & Pollack, A. (2014, September 22). Finding risks, not answers, in gene tests. *The New York Times*. Retrieved from http://www.nytimes.com

Harmon, A. (2006, April 12). Seeking ancestry in DNA ties uncovered by tests. *The New York Times*. Retrieved from http://www.nytimes.com

Harmon, A. (2007, November 11). In DNA era, new worries about prejudice. *The New York Times*. Retrieved from http://www.nytimes.com

Harmon, A. (2008, March 4). Gene map becomes a luxury item. *The New York Times*. Retrieved from http://www.nytimes.com

Harris, R. (2015, April 15). Personalizing cancer treatment with genetic tests can be tricky [Web log post]. National Public Radio. Retrieved from http://www.npr.org

Heshka, J. T., Palleschi, C., Howley, H., Wilson, B., & Wells, P. S. (2008). A systematic review of perceived risks, psychological and behavioral impacts of genetic testing. *Genetics in Medicine, 10*, 19–32. doi: 10.1097/GIM.0b013e31815f524f

Ishiguro, K. (2005). *Never let me go*. New York, NY: Knopf.

Johnson, K. (2015, April 4). "Gertie's babies," sold at birth, use DNA to unlock secret past. *The New York Times*. Retrieved from http://www.nytimes.com

Jolie, A. (2013, May 14). My medical choice. *The New York Times*. Retrieved from http://www.nytimes.com

Kolata, G. (2012, April 12). Study says DNA's power to predict illness is limited. *The New York Times*. Retrieved from http://www.nytimes.com

Kolata, G. (2014, December 28). In a new approach to fighting disease, helpful genetic mutations are sought. *The New York Times*. Retrieved from http://www.nytimes.com

Kolata, G. (2015, April 24). Chinese scientists edit genes of human embryos, raising concerns. *The New York Times*, p. A3. Retrieved from http://www.nytimes.com

Lerman, C., Croyle, R. T., Tercyak, K. P., & Hamann, H. (2002). Genetic testing: Psychological aspects and implications. *Journal of Consulting and Clinical Psychology, 70*(3), 784–797.

Lewin, T. (2000, July 21). Boom in gene testing raises questions on sharing results. *The New York Times*. Retrieved from http://www.nytimes.com

National Institutes of Health (2006, February). The future of genetic testing. *News in Health*. Retrieved from http://newsinhealth.nih.gov

Pear, R. (2008, January 18). Growth of genetic tests concerns federal panel. *The New York Times*. Retrieved from http://www.nytimes.com

Pear, R. (2015, January 24). Obama to request research funding for treatments tailored to patients' DNA. *The New York Times*. Retrieved from http://www.nytimes.com

Pinker, S. (2009, January 7). My genome, myself. *The New York Times*. Retrieved from http://www.nytimes.com

Pollack, A. (2013, November 25). F.D.A. orders genetic testing firm to stop selling DNA analysis service. *The New York Times*. Retrieved from http://www.nytimes.com

Rabin, R. C. (2013, November 26). In Israel, a push to screen for cancer gene leaves many conflicted. *The New York Times*. Retrieved from http://www.nytimes.com

Stein, R. (2015, April 21). Screening tests for breast cancer genes just got cheaper. National Public Radio. Retrieved from http://www.npr.org

10
The Microbiome

"Gut reaction," "gut feeling," and "have you got the guts for it?" take on new meanings now that scientists explore how our gut bacteria not only defend us against disease-causing microbes (Courage, 2015; McGreevey, 2014; Yong, 2014) but also may affect who we are. Mobilizing microbes for good causes is spawning industries. Microbiome banks, similar to blood banks, provide specimens for fecal transplants (Belluck, 2014; Courage, 2014a). Full understanding is lacking, treatments could have unintended consequences (Gallagher, 2015), and the FDA is unsure how to categorize these new options (Shaffer, 2014). Large-scale efforts to collect and catalog personal microbiomes are underway in at least three countries (Stein, 2013). Although such databases raise privacy issues (Shaw, 2014), they may help researchers develop "personalized medicine."

Caring for our own microbiomes is used to promote natural childbirth, higher-fiber diets, restricted use of antibiotics, less sterile home environments (Murphy, 2015), and less stringent cleansers (Doucleff, 2015; Eakin, 2014; Scott; 2014; Shaw, 2014). Others warn that we must not "romanticize our relationship" with microbiomes because the same microbe can be helpful or harmful depending on circumstances (Yong, 2014). But the idea that we may be composed of a community of life forms, not all human, stimulates a host of ethical issues. It's a new frontier inside us. What could this mean to our understandings of who – and what – we are?

Am I a "We"?

with Ria Citrin

Microbiome research followed the success of The Human Genome Project (National Institutes of Health [NIH], 2014). Microbiomes may help with obesity, diabetes, depression, autism, and other ailments (Hsiao, 2013). But not until The Human Microbiome Project launched did we appreciate that humans cohabitate with – and depend on – bacteria, viruses, fungi, and protozoa. They are all over and in our bodies, most notably the intestines (Yang, 2012). Microbial cells outnumber human cells ten to one (University of Utah Health Sciences, 2014).

For some, this surprising body-composition ratio may raise concerns about "germs." It may be particularly troubling to sufferers of obsessive-compulsive disorder, whose prevalence is on the rise (Veale, 2014). Since the 1800s, when microbes were linked to infections and contagions, they have been considered primarily an enemy. If our microbiome is *integrated* with our humanity, then no amount of medications can eradicate "germs" in a healthy way. Our microbiome helps us digest food, absorb nutrients, metabolize medicines, support our immune system (Courage, 2014a; Doucleff, 2014) – and, counterintuitively, can keep us from stinking (Rutsch, 2015; Scott, 2014).

For others, the notion of a microbiome – that we are "at one" with other living beings and should not be at war with them – is welcomed. The current backlash against antibiotics and antibacterial products supports this perspective (Stromberg, 2014). We cannot control the presence of microbes in the body, and perhaps there's no sense to try, because they're necessary and helpful (Courage, 2015). The effects microbiomes may have on us should be qualified by the context (Yong, 2014).

Besides lacking consensus of whether microbes are good or bad for health and medicine, emerging microbiome ideas pose ethical quandaries related to personal control, responsibility, and individuality.

Who's in control?

Control addresses restraint over another. Stoic philosophers considered control as immediate, something that can only occur in a present situation (Evans, 2014). One ethical issue, then, becomes: Are "we" humans in control of our bodies, or are "they"?

A diminished sense of human control isn't necessarily negative. It may be good to recognize the contributions of other life forms to our well-being (Yang, 2012) – not only gut or skin microbes, but plants and animals via food, and insects like bees for pollination or honey. The endosymbiosis theory suggests that mitochondria inside our cells, which produce the energy needed for our bodies, are bacteria absorbed during our evolutionary history (University of California Museum of Paleontology, 2012). Why would giving credit to bacteria for our digestion or immune systems not be beneficial? If people saw microbes as helpful, reduced hysteria around germs might help with cooperation in health initiatives (Doucleff, 2014).

Realizing we may have less control over our own bodies can be unnerving. Will microbes treat us well or not? An interesting comparison is to a performance art piece in which the artist placed 72 objects in a room and allowed audience members to use the objects for or against her in any way they pleased. Some of the objects were harmless, whereas others were potentially violent. By the end of the piece, the artist's clothes were cut to rags, and one audience member had pointed a loaded gun at her head (Lasane, 2014). This art demonstrates how lack of control can be frightful.

The lack of control associated with microbiomes, however, may be scarier. The art was about relinquishing control to others, whereas consideration of microbiomes is about realizing we never had control in the first place. The artist made a *choice* to be passive to the audience, which involves risk. But humans don't have the option to be outside the control of our microbiomes. If our microbiomes go awry – as can happen if they get out of balance or are maltreated by our diets, habits, or even our pets (Doucleff, 2015; Gallagher, 2015; Zimmer, 2006) – they have the potential to point a biological "loaded gun" at us without our discretion. While it's impossible to live without microbiomes, the idea that they control us can change our ideas of agency and autonomy; we are *dependent*.

Where does responsibility lie?

If we aren't in control, are we responsible? Whereas control addresses restraints imposed on a situation, responsibility addresses capability to fulfill an obligation and culpability if we don't. Microbiome transplants have treated mice with chronic illnesses such as obesity, diabetes, and even depression and anxiety (Hsiao, 2013). Humans have been cured of at least one antibiotic-resistant intestinal infection in this way (Eakin, 2014). If microbes are responsible for diseases, and can become responsible for cures, what else might they be responsible for?

What about human behavior? Is a new legal defense in the works: "My microbiome made me do it"? Even if the law does not recognize this argument, this reasoning has consequences for ethics. If microbes can control behavior, our conception that people act independently may be unfounded.

This notion is not as far-fetched as it may first seem (Zimmer, 2014). Studies show how some parasites make mice less afraid of cats so the parasite is ingested by cats to reproduce (Zimmer, 2006). Our microbiomes may manipulate how we eat (Alcock, Maley & Aktipis, 2014), how we feel or handle stress (Stilling, Dinan & Cryan, 2014), and how we make friends (Zimmer, 2014).

These findings are problematic because we may use our microbiome as an excuse for unhealthy habits and bad behaviors. Self-help might become less compelling. Why would people with Type 2 diabetes change their diets and lifestyles if it's really their guts that are at fault, or if the illness can be cured more simply with a fecal transplant from someone without diabetes?

Conversely, these findings are promising because, perhaps, the proper care and feeding of our microbiomes might make us happier. Not only could depression and anxiety abate, but also the stigma associated with them. The understanding that sufferers might not be at fault for their illness would be comforting (Oxford Brookes University, 2012). Yet, on a more troubling note, we might end up lacking any feelings at all. "Bad" feelings wouldn't exist, which would make it impossible to define "good" feelings.

To take this argument a step further, if all we need is the right concoction of microbes, perhaps we could become "perfectly" happy and healthy. Some billionaires invest considerable money in doctors, personal trainers, specialized diets, and longevity research (Alsever, 2013). Instead of investing our own efforts to strive for our health or lifestyle goals, we buy a better microbiome. Perhaps microbiotic therapy would take less money, so health would be achievable by people with lower economic means. Let's say we figure out the ideal balance of amounts and types of cells within our bodies. People would live much longer, perhaps forever. Not only could this lead to overpopulation, it also might change our value of life. If everyone has perfect health, and no emotional ups and downs, life might lose meaning.

Now suppose the opposite: humans have *no* control over their "human nature" at all. From one philosophical perspective, human nature is primal and social rules are learned (Rousseau, 2014). What would happen to manners, social conventions, and institutions built upon human nature? Many legal systems are predicated on the notion that a person is a responsible entity. Efficient organization might become problematic unless our microbiomes could get "in sync" with each other – society wouldn't work if our microbiomes led to us "belly-aching" at each other. Perhaps both betrayal and honesty increase as people behave openly on their immediate "gut instincts." Many movies and books explore what would happen if a character couldn't lie (TV Tropes, n.d.). What would change in how we cooperated, coordinated, and communicated if we no longer thought that we were interacting with a "person" but rather a bacterial colony?

What happens to individuality and autonomy?

If we consider our microbiomes responsible for how we behave, it is not that big a step to suggest they compose our "selves." Existential psychology proposes that much of the human condition struggles with acceptance of the uncertain by making meaning, and that uncertainty relates to lack of agency (Existential-Humanistic Institute, n.d.; Olivares, 2010).

The progress of microbiome research may lead people to question "who am I?" in two ways. First, am I a "who" – a self with subjective authority? This question focuses on autonomy, a state of being able to handle our own affairs. As the discussion of responsibility above suggests, we may not be authors of our own lives. Second, am I an "I" – an undivided whole? This question focuses on individuality, a word based on the same root "div" as the word "divide." An individual is an inseparable whole. If microbial cells outnumber human cells, and those microbes have their own agenda, then are humans indivisible? Symbiosis does not equal sameness or identification. We are not our microbes, but are we simply vessels for them?

Then again, our microbiomes – if we develop technology to manipulate *them* through diet, pharmaceuticals, or some other yet-to-be-invented treatment – may become another tool for us to further individuate from each other. We could take our "uniquenesses" to the extreme. We could custom design ourselves to our own beauty ideals or in other potentially advantageous ways. Perhaps we would be able to change our appearance constantly. A dinner date might return from the bathroom and be unrecognizable, but no one would be shocked. This reality is startlingly close – plastic surgery and other methods to alter appearance are possible for those with the money, resources, and desire. In South Korea, which has the highest rates of plastic surgery in

Where do "we" go from here?

Prior scientific innovations led to questioning accepted beliefs. Charles Darwin faced a difficult decision whether to publish his theory of evolution because he perceived it could threaten prevalent religious ideas of his time and culture (Schroeder, n.d.). Microbiomes have the potential to cause similar pervasive uncertainty and, perhaps, a paradigm shift in how we see ourselves and our relation to the universe. Whereas Darwin's theory addressed humans' relationship to the "larger" world, the microbiome focuses on humans' relationship to the "smaller" world.

Or perhaps these ethical issues are not germane. Let's stipulate that we have no free will and our microbiomes just tell us what to do. If we don't believe that or act on that belief, is it "true"? If, as social constructivists propose, our social reality is made up in our shared ideas (Vygotsky, 1978), and some of our "ideas" come from our gut bacteria, then will we be able to tell the difference between the role of the human brain cells and the microbial cells? If perception or belief that we have control is strong enough to overcome indicators that we have little – or perhaps zero – control, maybe the ethical implications associated with microbiomes aren't too pernicious.

The ethics of microbiomes is not about control per se, but the *sense* that humans have control. The shorter-term effects of microbiomes as cohabitants in our bodies may be relatively benign. However, graver implications may arise over time as we give up our sense of agency, responsibility, free will – and perhaps our sense that we even exist – to these microorganisms. Does this scenario have to be settled as an either-or solution? Or are there other possibilities, such as humans and microorganisms respecting each other's roles in our mutual well-being? In the same way humans came together to live in societies, can we and other life forms build ways of thinking that allow for cross-species "institutions"?

Further exploration

1. What would society look like if we discriminated based on the microbial fingerprint?
2. What would happen if autonomy disintegrates?

References

Alcock, J., Maley, C. C., & Aktipis, C. A. (2014, October). Is eating behavior manipulated by the gastrointestinal microbiota? *BioEssays, 36*(10), 940–949. Retrieved from http://onlinelibrary.wiley.com/doi/10.1002/bies.201400071

Alsever, J. (2013, April). 5 billionaires who want to live forever. *Fortune.* Retrieved from http://fortune.com

Belluck, P. (2014, October 11). A promising pill, not so hard to swallow. *The New York Times.* Retrieved from http://www.nytimes.com

Courage, K. H. (2014a, December 12). Poo and you: A journey into the guts of a microbiome [Shots web log]. National Public Radio. Retrieved from http://www.npr.org

Courage, K. H. (2014b, December 17). Behind the scenes at the lab that fingerprints microbiomes [Shots web log]. National Public Radio. Retrieved from http://www.npr.org

Courage, K. H. (2015, January 18). One scientist's race to help microbes help you [Shots web log]. National Public Radio. Retrieved from http://www.npr.org

Doucleff, M. (2014, November 14). How bacteria in the gut help fight off viruses. National Public Radio. Retrieved from http://www.npr.org

Doucleff, M. (2015, April 21). How modern life depletes our gut microbes [Goats and Soda web log]. National Public Radio. Retrieved from http://www.npr.org

Eakin, E. (2014, December 1). The excrement experiment. *The New Yorker*, pp. 64–71.

Evans, J. (2014). How people follow Epicurus' philosophy today. Retrieved from http://philosophyforlife.org

Existential-Humanistic Institute (n.d.). About existential therapy. Retrieved from http://www.ehinstitute.org

Gallagher, J. (2015, February 7). Woman's stool transplant leads to "tremendous weight gain." BBC Health. Retrieved from http://www.bbc.com

Hsiao, E. (2013, February 8). Mind-altering microbes: how the microbiome affects brain and behavior: Elaine Hsiao at TEDxCaltech [Video file]. Retrieved from https://www.youtube.com

Lasane, A. (2014, January). Marina Abramovic reflects on her important "Rhythm 0" piece from 1974. *Complex.* Retrieved from http://www.complex.com

McGreevey, S. (2014, May 14). Keeping healthy bacteria happy. Retrieved from hms.harvard.edu/news

Murphy, K. (2015, May 9). Invite some germs to dinner. *The New York Times.* Retrieved from http://www.nytimes.com

National Institutes of Health. (2014). Human Microbiome Project overview. Retrieved from https://commonfund.nih.gov

Olivares, O. J. (2010). Meaning making, uncertainty reduction, and the functions of autobiographical memory: A relational framework. *Review of General Psychology, 14*(3), 204–211.

Oxford Brookes University. (2012). Overcoming depression. Retrieved from https://www.brookes.ac.uk

Roffee, B. (2014, April). Plastic surgery is so extreme in South Korea that people need new IDs. *Ryot.* Retrieved from http://www.ryot.org

Rousseau, J. J. (2014). *Discourse on the origin and the foundations of inequality among men* (I. Johnston, Trans.). Adelaide, Australia: The University of Adelaide. Retrieved from https://ebooks.adelaide.edu.au (Original work published 1754).

Rutsch, P. (2015, March 31). Meet the bacteria that make a stink in your pits [Health web log]. National Public Radio. Retrieved from http://www.npr.org

Schroeder, D. (n.d.). Evolutionary ethics. In J. Fielser & B. Dowden (Eds.), *Internet Encyclopedia of Philosophy*. Retrieved from http://www.iep.utm.edu

Scott, J. (2014, May 25). My no-soap, no-shampoo, bacteria-rich hygiene experiment. *The New York Times Magazine*, p. MM28. Retrieved from http://www.nytimes.com

Shaffer, A. (2014, September). When feces is the best medicine. *The Atlantic*. Retrieved from http://www.theatlantic.com

Shaw, J. (2014, March-April). Why "big data" are a big deal. *Harvard Magazine*, pp. 30–35, 74–75.

Stein, R. (2013, November 4). Getting your microbes analyzed raises big privacy issues. National Public Radio. Retrieved from http://www.npr.org

Stilling, R. M., Dinan, T. G., & Cryan, J. F. (2014, January). Microbial genes, brain & behavior – epigenetic regulation of the gut-brain axis. *Genes, Brain and Behavior, 13*(1), 69–86. doi: 10.1111/gbb.12109

Stromberg, J. (2014, January 4). Five reasons why you should probably stop using antibacterial soap. *Smithsonian Magazine*. Retrieved from http://www.easybib.com

TV Tropes. (n.d.). Cannot tell a lie. Retrieved from http://tvtropes.org

University of California Museum of Paleontology. (2012). Cells within cells: An extraordinary claim with extraordinary evidence. Retrieved from http://undsci.berkeley.edu/lessons/pdfs/endosymbiosis.pdf

University of Utah Health Sciences. (2014). The human microbiome. Retrieved from http://learn.genetics.utah.edu

Veale, D. (2014, April 4). Obsessive-compulsive disorder. *British Medical Journal*, p. 348. doi: 10.1136/bmj.g2183

Vygotsky, L. S. (1978). *Mind in society*. Cambridge, MA: Harvard University Press.

Yang, J. (2012, July 16). The Human Microbiome Project: Extending the definition of what constitutes a human. National Human Genome Research Institute. Retrieved from http://www.genome.gov

Yong, E. (2014, November 1). There is no "healthy" microbiome. *The New York Times*. Retrieved from http://www.nytimes.com

Zimmer, C. (2006, June 20). A common parasite reveals its strongest asset: Stealth. *The New York Times*. Retrieved from http://www.nytimes.com

Zimmer, C. (2014, August 14). Our microbiome may be looking out for itself. *The New York Times*. Retrieved from http://www.nytimes.com

11
Stem Cell Therapy

Stem cells can rebuild broken body organs from "original" cells, and some scientists see relieving suffering as a "moral duty" (Gallagher, 2015). But if stem cell therapies go awry, they could breach medicine's Hippocratic Oath of "do no harm" far worse than current treatments. Anxieties abound regarding what could go wrong, or what unanticipated consequences could develop, even if therapies go well.

Although the European Union has approved its first therapeutic use of stem cells (Gallagher, 2014a), and many clinical trials are underway (Weintraub, 2014), as of 2015, the United States prohibits use of federal funds to match stem cells to patients' DNA for clinical use (Reuters, 2014). However, scientists continue to develop potential stem cell applications in medicine, such as for stroke (Mundasad, 2014), degenerative eye diseases (Pollack, 2014), insulin-producing cells to counteract diabetes (Stein, 2014b), and cystic fibrosis (Regalado, 2015). Stem cell sources have expanded from embryos and fetuses to adult tissue (Harris, 2014). But, some "breakthroughs" have been debunked (Doucleff, 2014; Gallagher, 2014b), so much uncertainty continues.

There are worries that researchers, doctors, and individuals may shift stem cell use from cures to enhancement, spawning a "body-by-design" industry (Regalado, 2015; Stein, 2015). Even more concerning is stem cell manufacturing of a "new and improved" future species, launching Humans 2.0. The United Nations has lobbied countries to ban such uses (Reuters, 2014), and the European Union considers it a "crime against human dignity" (Regalado, 2015), but the United States has not banned this "germline engineering" of traits that can be passed on to future generations (Regalado, 2015). Some scientists call for a moratorium (Stein, 2015), whereas others want to clarify its benefits and harms (Stein, 2014a; Wade, 2015). Still, at least one scientific team tried to address these issues empirically (Kolata, 2015). Will such developments expand or contract the notion of family, individuality, and the value of life?

Biological Reboot
with Sarah Schnur

Stem cells are unspecialized cells that can develop into any type of cell – a heart, lung, skin, or muscle cell, for example. They are remarkable because they can replenish healthy tissue as well as replace diseased tissue. Thus, they might be used in a wide array of therapeutic applications, from testing new drugs, to replacing damaged heart ventricles, to re-growing a liver rather than waiting for a donor's liver to become available for transplant (National Institutes of Health, 2009). Stem cells initially came from embryos and fetuses, but now can also be derived from "reprogrammed" adult tissue (National Institutes of Health, 2009). Stem cell research started in the 1950s, organized in the 1960s, and was banned in the 1970s. In the 2010s, it is allowed under strict regulation (Stem Cell History, 2013).

The multipotentiality and sources of stem cells give rise to numerous ethical dilemmas not only in medical research and therapy but beyond into our ideas about when tissues fall under the protection of human rights, how therapeutic benefits are distributed, and accelerated evolution of the human species. As government agencies aim to mediate medical research and therapy opportunities, the considerations of religious concerns, treatment beneficiaries, and society's collective interests must be addressed.

Potential persons as raw material

Stem cells can be collected from aborted fetuses (Wertz, 2002). But is it morally justifiable to destroy fetal tissue – or *create* fetal tissue – to collect stem cells? Some scientists view fetal tissue as ideal, especially if the fetus would have been aborted regardless. Researchers also would like to use the leftover fertilized cells from ineffective in vitro fertilization procedures. This view is akin to repurposing living tissue whose original purpose was thwarted.

However, the government is not keen to provide incentives for women to abort pregnancies, such as by researchers starting to pay women to get pregnant or have an abortion just to produce stem cells (Wertz, 2002). Over the longer term, turning babies into pay might lead to emotional detachment between mothers and their future babies if they became used to receiving payment for pregnancies. Fertile adults may take a more active role in addressing their own or loved ones' health problems by purposefully planning an abortion to provide stem cells.

Others fear that abortions could become justified because people feel "some good would come of" using the stem cells therapeutically (Wertz, 2002). But it is possible that patients who benefit from stem cell therapies may feel guilt or anxiety from the benefit of fetal tissue, creating a revenge effect (Tenner, 1996) that leads to poorer prognoses for these patients.

A third view, that of religious groups, advocates that every fertilized cell needs to be protected, and intentional use of abortion for instrumental purposes equals premeditated murder (Wertz, 2002). Religious conservatives also object to embryonic tissue created in a scientific laboratory with the intent to destroy it for research purposes. From conception, these tissues are considered potential persons. Yet, conceiving them as biological material for research or therapies privileges other persons over the not-yet-born, who have no say in the matter.

The extreme case is people-as-body-parts, created in the lab solely for the health and welfare of others (see novel by Ishiguro, 2005). These babies never had parents in the social sense, or an anticipated lifespan to live, learn, love. Why stop only at stem cells? A "slippery slope" of possibilities could arise as these tissues never cohere into an actual, living baby, but stay as separated tissues for experimentation. Where is the well-drawn line for them to be considered "human subjects" in terms of research ethics and human rights?

Winning the lottery of life

Among the many sick individuals who might benefit from stem cell therapy, who receives treatment? And who decides who those lucky patients are? Stem cell therapy can be expensive (Wertz, 2002). Will stem cells simply go to the highest bidder? This economic perspective sets up an inequality in that those with the most are more likely to live longer to acquire even more. Although it is not a novel concept that the wealthy and powerful enjoy more access to resources – including medical care – than the less well endowed, it is new that stem cell therapies give them access to the origins of human life.

Stem cells are the foundations from which each and every cell in a human body develops. As technologies develop that enhance researchers' and doctors' abilities to manipulate stem cells into any type of cell, the possibility may emerge to transform the aspiration of "be all I can be" into "become whatever I want to be." Those who can afford it could regenerate parts of their bodies on whim – not just cosmetically but physiologically. Want intestines that do not get upset so easily?

No problem...a new digestive tract can be manufactured from stem cells. Want a perkier personality and an easier time making decisions? Perhaps a revised prefrontal cortex grown from stem cells might make that possible someday.

Stem cell therapies eventually could become high-tech body sculpting procedures – like an extreme form of plastic surgery not only for visible but also internal parts of our bodies. Forget silicone implants. Say good-bye to cellulite and dentures. Have new breasts, gluteal muscles, skin, and teeth generated from stem cells. If restrictions on stem cell production are removed, stem cells could become less expensive and abundant. How much personal or societal resources should go toward body redesigns? Should they be covered by insurance or government health programs?

When and how should the government step in with financial support (Fulton, Felton, Pareja, Potischman, & Sceffler, 2009)? What criteria should officials use to distribute stem cells? Because stem cells can become any type of tissue, their deployment is not limited in the way organ donations are. When a donor liver becomes available, it goes to a recipient high on the in-need list for that particular organ who is a good match on other criteria. Only patients in need of livers are considered. With stem cells, an additional consideration is what type of tissue is more valuable? The stem cells could simultaneously be demanded for a new heart due to a congenital defect, or new skin due to a severe burn, or new eyeballs for a blind person. The possibilities are nearly limitless, which makes the choices even harder.

Another perspective on who holds the power to decide is familial. A child may have a health problem that could be addressed by stem cell therapy. Does the child or the parents make the call? What if the parents want to regenerate a part of their child they consider to be less than ideal – not muscular enough for sports, perhaps? What if the parents have the money for the therapy but choose not to provide it for their child? If the stem cell therapy works, who gets credit? If it doesn't work, who is blamed?

Changing life from precious to artifice

Stem cell therapies come close to genetic manipulation of human germ-lines, the DNA that are found in every cell of a particular body and that would be passed on to the next generation. If scientists are able to create fetal tissue, why not create a fetus – or eggs and sperm – leading to specific traits? Not only parents of a particular baby but leaders of a nation or ethnicity or race might decide that particular traits are required and "program them in" to future generations by manipulating the genes of germline stem cells. A new high-tech eugenics movement starts.

Parents or nations may ask whether the trouble, expense, and responsibility of children naturally conceived, born, and raised are worth it versus children manufactured via a more scientifically controlled process of physical – and perhaps intellectual and emotional – development. Society may come to rely on the laboratory as the sole source of children. The concept and label of parent eventually has no meaning. The social functions of families and siblings become extinct. It may become confusing to define oneself if the child born is a mixture of natural and manipulated genetic traits. Children born naturally into the human population could be regarded as of less value or status. If enough parents resorted to this method of having children, after many generations, women may lose the ability to become pregnant. The "ties that bind" us now, and have for millennia, become unwound.

If this method of stem cell creation is successful for humans, why not for other life? It may become easy to create life, therefore making it replaceable and devalued. People could ask for specific breeds – or perhaps interbreeds – of pets (which would turn dogs called "labs" into a joke). Scientists could design food from stem cells – beets with less greens and more root, grains with more wheat and less chaff. Product testing on animals could recommence. Scientists may be able to create animals with specific traits they want to test against. For example, if a scientist wanted to study the effects of a synthetic food on depressed dogs, they could genetically manipulate dogs to have lower serotonin levels. Stem cell productions could replenish endangered species – or design new species. More hunting may be allowed because it is possible for scientists to create more prey. Manufacture makes conservation unnecessary. We become sustainable by making more, not using less.

Just as stem cell therapies can spawn branches of possible types of physical tissues, so too can ethical questions surrounding stem cell usage bring into being a host of both exciting and troubling scenarios. Will the ability to produce life from cells that contain all possibility – that can become anything – make life in general, and our lives in particular, more or less valuable? Will we make more careful or more careless decisions about our current, natural bodies? Will the opportunity to replace any part of ourselves through controlled bio-manufacturing help us treasure who we are or start us chasing bio-fads and bio-fashions? Will stem cells converge on some new medicalized standard for status-seeking or accentuate a broader, more enlightened view of diversity?

Further exploration

1. If stem cells provide the ability to perennially replace any living thing – plants, animals, people – what happens to the value of life that already exists?
2. What are societal implications of some people gaining an intellectual, athletic, or other advantage due to stem cell therapy?

References

Doucleff, M. (2014, April 1). Fraud found in study claiming fast, easy stem cells. National Public Radio. Retrieved from http://www.npr.org

Fulton, B. D., Felton, M. C., Pareja, C., Potischman, A., & Sceffler, R. M. (2009). *Coverage, cost-control mechanisms, and financial risk-sharing alternatives of high-cost health care technologies.* Berkeley, CA: California Institute for Regenerative Medicine.

Gallagher, J. (2014a, December 19). Stem cells: First therapy approved by EU. British Broadcasting Corporation. Retrieved from http://www.bbc.com

Gallagher, J. (2014b, March 14). Stem cell "breakthrough data inappropriately handled." British Broadcasting Corporation. Retrieved from http://www.bbc.com

Gallagher, J. (2015, May 13). Embryo engineering a moral duty, says top scientist. British Broadcasting Corporation. Retrieved from http://www.bbc.com

Harris, R. (2014, April 17). First embryonic stem cells cloned from a man's skin. National Public Radio. Retrieved from http://www.npr.org

Ishiguro, K. (2005). *Never let me go.* New York, NY: Knopf.

Kolata, G. (2015, April 24). Chinese scientists edit genes of human embryos, raising concerns. *The New York Times*, p. A3. Retrieved from http://www.nytimes.com

Mundasad, S. (2014, August 9). Stem cells show promise in stroke recovery. British Broadcasting Corporation. Retrieved from http://www.bbc.com

National Institutes of Health. (2009). *Stem cell information.* Retrieved from http://stemcells.nih.gov/Pages/Default.aspx

Pollack, A. (2014, October 14). Study backs use of stem cells in retinas. *The New York Times.* Retrieved from http://www.nytimes.com

Regalado, A. (2015, March 5). Engineering the perfect baby. *MIT Technology Review.* Retrieved from http://www.technologyreview.com

Reuters, (2014, April 17). Scientists report advance in "therapeutic cloning." *The New York Times.* Retrieved from http://www.nytimes.com

Stein, R. (2014a, February 26). Scientists question safety of genetically altering human eggs. National Public Radio. Retrieved from http://www.npr.org

Stein, R. (2014b, October 9). Scientists coax human embryonic stem cells into making insulin [Shots web log]. National Public Radio. Retrieved from http://www.npr.org

Stein, R. (2015, March 20). Scientists urge temporary moratorium on human genome edits [Shots web log]. National Public Radio. Retrieved from http://www.npr.org

Stem Cell History. (2013, January 1). Stem cell research timeline. [Web log post]. Retrieved from: http://www.stemcellhistory.com

Tenner, E. (1996). *Why things bite back: Technology and the revenge of unintended consequences*. Cambridge, MA: MIT Press.

Wade, N. (2015, March 20). Scientists seek ban on method of editing the human genome. *The New York Times*, p. A1. Retrieved from http://www.nytimes.com

Weintraub, K. (2014, September 15). The trials of stem cell therapy. *The New York Times*. Retrieved from http://www.nytimes.com

Wertz, D.C. (2002, June). Embryo and stem cell research in the United States: History and politics. *Gene Therapy, 9*, 674–678. doi: 10.1038/sj/gt/3301744

12
Fortified Junk Food

The word "vitamin" comes from the same root as "vital," meaning life. Vitamins are necessary for the proper functioning of our bodies (Zimmer, 2013). Since we realized they could eradicate many diseases, like scurvy or rickets, vitamin use has grown tremendously. Governments passed laws to fortify processed food staples – like cereals, flour, and mixes – to provide everyone with at least some vitamins. But studies show that we can eat too much of a good thing (Bjelakovic, Nikolova, Gluud, Simonetti, & Gluud, 2007).

An abundance of reformulated products – usually considered junk foods – have come on the market labeled as "high fiber" or "20% of daily Vitamin C" or "made with whole grains" (e.g., Choi, 2014; Eng, 2015). Although some critics of the whole food movement consider fortified junk food helpful to counteract obesity by "tricking" people into losing weight (Freedman, 2013), many worry that these fortified junk foods will create even more confusion about nutrition. These processed products will obfuscate what "healthy" means, especially since food companies are now allowed more self-monitoring (Quinn, 2015). This case presents a fictional, and somewhat satirical, scenario outlining the types of ethical issues and challenges a food company might face, including situations where standard marketing or business practices may create ethical quandaries. Do some vitamins really make food higher quality?

Too Much of a Good Thing?

with Lilia Juarez-Kim

CONFIDENTIAL INTERNAL MEMO
TO: Executive Committee, Random Food Company
FROM: Lilia Juarez, Consultant
DATE: May 15, 2015
RE: Moving forward with fortified junk food

It has come to my attention that you plan to fortify all the company's snack and convenience food products. Fortification incorporates vitamins, minerals, protein, and fiber into processed foods (FDA, 2010; Schmeck, 1974). Historically, fortification programs were mandatory to replenish nutrients lost in processing, such as in breakfast cereals and flour. They worked well and spread. Some became overseen by the United Nations to improve nutrition worldwide (Codex Alimentarious Commission, 1991). Your plan to fortify your snack line is not required by law, but is a voluntary undertaking (Sacco & Tarasuk, 2009).

As I have learned from observing your company, fortification spells opportunity. Increasingly over the 20th century, Americans have eaten more processed foods that are convenient and require little cooking know-how other than to "just follow the directions" on the container (Food Additives, 1959). But people want to feel like they are being good to themselves, and government and "whole foods" activists have become better at convincing consumers to eat healthier. Recently, you and your competitors realized that *any* food product might be viewed favorably as a "health food" with enough fortification. This started a race to infuse nutrients into candies, carbonated beverages, sweet or salty snacks, and convenience foods.

Since your company encourages innovation, this memo outlines important ethical considerations as you move forward.

1. **Some countries discourage the fortification of junk food** (Food Fortification, 2010) because it is believed to mislead the consumer (Codex Alimentarious Commission, 1991). In the United States, your biggest market, the Food and Drug Administration (FDA) states that fortification is appropriate only when there is a justified need because of nutrient loss in processing or to improve nutrient balance (FDA, 2013). Therefore, fortifying junk food may not be appropriate (Backstrand, 2002). But the FDA lacks enforcement power and is still waiting for Congressional authorization to conduct a study on the

effects of fortified junk food (FDA, 2013). With this bureaucratic hold-up, it is even more important for you to consider the ethical implications since you are expected to conduct your own studies and monitor yourself (Kindy, 2014).
2. **It is important to consider what vitamins Americans actually** *need* (FDA, 2010). As you are aware, your competitors have launched cookies fortified with vitamin D (Sakimura, 2013) and soft drinks with added B12 and magnesium (Schleicher, 2007). In fact, you are entering this market late as fortified junk food in grocery stores has proliferated, and bloggers are starting to take notice (Nestle, 2013; Sakimura, 2013). After considerable review of these online conversations, they focus on a lack of transparency about what nutrients companies plan to use. For instance, vitamin A may not be a good option because Americans are not deficient in it (Miller, 2012), and too much vitamin A can lead to health problems and even death (Bjelakovic et al., 2007).
3. **It would be prudent to take a "big picture" on nutrients.** More is not always better. Choosing random vitamins can create health issues for the consumer, and a sick consumer becomes a nonconsumer of your products in the future. Let's assume you are successful in capturing a large market by providing a well-rounded selection of snacks and convenience foods. Your brand skyrockets, and you get a loyal following of snackers. You may need to make sure they don't overdose if they eat five or more Lotta-a-Chocs with calcium in a day. If the fiber in Cheesy Pizzees inhibits the absorption of the B vitamins infused in Near Juz, you may not want to market these products as a combo. Plus, it would be wise to consider how adding vitamins or fiber does not alleviate possible health problems from other ingredients in the snack, like sugar substitutes (Aubrey, 2013).
4. **Fortification may confuse consumers since they may replace well-balanced meals with vitamin-infused snacks** (Sacco & Tarasuk, 2009). In the short term, it may seem that this is exactly what you want to happen since that would increase market share. But that could backfire if people get tired of your snacks from eating them all the time. On the other hand, I do see a long-term trend toward convenience foods as consumers don't realize they lose cooking skills by eating only processed foods. Individuals too stressed or busy to think about what they eat, or who prize convenience too highly to take time for anything other than microwavable dinners (Kindy,

2014), may never come in contact with fresh foods that they would have to prepare themselves.

This situation may be good for your company and your products because consumers would have no basis of comparison: they will forget what it is like to eat a fresh peach instead of your Peach-Ums! with extra Vitamin C, or barbecue a salmon fillet rather than microwave your Something's Fishy Cakes with fiber. Best of all, if you can train the taste buds of children that your products are what "real food" is, then you won't have to work so hard hiding the vitamin and mineral aftertaste in iron-infused Fluffies. The increased advertising of fortified junk food may create a belief that it is 100% healthy, so people will feel good about eating more of your products. They may never make the connection between their chronic disease and junk food because they think vitamin-packed OB-So-Good nougat is a healthy choice.

5. **Natural or synthetic vitamins?** As you know, some vitamins don't survive processing (Miller, 2012), especially if they derive naturally from plant and animal sources. If consumers come to depend on products as healthy sources of these nutrients, you may contribute to a vitamin deficiency situation because they think your product has a vitamin that didn't survive the processing. For example, a pregnant woman drinks a BGood Avocado Smoothie every day. Your original formulation uses some real avocados, but their vitamin B5 is lost in processing. She thinks she is getting B5, but she's not, leaving her with an unknown potential deficiency that may put her unborn child at risk for abnormal body development. That would be bad for business.

Synthetic vitamins pose different risks. They can mask that some foods have no real underlying nutritional value (Price, 2015). Furthermore, too much processed vitamins could create a new chronic disease, and you don't want to be associated with that. There are many unknowns about long-term effects of synthetic vitamins, not only on individuals who consume them, but also effects on reproductive health, or effects of the chemicals on workers or released into communities from their manufacturing process. You might consider these risks more closely.

6. **Watch out for enticing marketing strategies that may backfire.** Don't try to be too clever with artificial representations of your foods (Rutsch, 2015b), or claims that are too outlandish and lead to consumer mistrust (Noguchi, 2015) or, even worse, a letter

from the FDA (Rutsch, 2015a). You provide individual serving sizes and consumers may think you are tailoring the healthfulness of a particular product to their own unique nutritional needs, despite all the different diets and food allergies today. Although people eat more – and more junk food – when they are bored (Canetti, Bachar & Berry, 2002), food can also be a social experience (Druckerman, 2015).

7. **Children and adolescents can be your best ally or worst nightmare.** Fortified junk food may bring your products back to school vending machines and cafeteria lunches (Eng, 2015; Noguchi, 2015). One scenario is that the prolonged and consistent consumption of sugar, fat, and the vitamins may potentially affect children's and adolescents' brain chemistry, influencing their overall learning, which lowers average American intelligence across future generations. Children may not understand nutrition and may consume more fortified junk food than any other group. Children's and adolescents' bodies need different amounts of vitamins or different vitamins all together. Fortified junk food may contain a standard amount of vitamins that is adequate for adults; however, the same amount in children may poison vital organs. You would not want to be responsible for that.

8. **Don't create "food deserts" in poor communities.** Making fortified junk food easily available in areas where fresh food is lacking could lead to a serious malnutrition problem, especially in their children. In some extreme cases, it may even replace protein sources because parents may not be able to afford meat as they work hard to pay for housing and bills. Malnutrition in future generations could create disadvantages in learning, lowering educational attainment, creating a wider jobs gap, into an unfortunate downward spiral from over-dependence on inexpensive fortified junk food.

9. **Consider impact on the natural environment.** Resources used to make junk food, which is then fortified to seem healthier, create costs to the environment. Vitamin extraction from natural sources increases demand for land and water to grow the required plants. This increased use may be particularly egregious if, as mentioned above, the vitamins don't even make it into the final product. The land might be put to better use for other crops.

Further exploration

1. Choose a role (CEO, parent, worker, or consumer) and propose an argument – from your role's perspective but also considering the other roles – whether junk food fortification should be extended worldwide and why or why not.
2. Imagine a society that no longer produces food naturally, but only through manmade, chemical manipulation. How does that change the meanings of "meal" or "food" or "nutrition"? What would "junk food" mean in a chemical-only world?

References

Aubrey, A. (2013, July 10). Do diet drinks mess up metabolisms? [The Salt web log]. National Public Radio. Retrieved from http://www.npr.org

Backstrand, J. R. (2002). The history of food fortification in the United States: A public health perspective. *Nutrition Reviews, 60*(1), 15–26. Retrieved from http://globalseminarhealth.wdfiles.com/local--files/nutrition/Backstrand.pdf

Bjelakovic, G., Nikolova, D., Gluud, L. L., Simonetti, R. G., & Gluud, C. (2007). Mortality in randomized trials of antioxidant supplements for primary and secondary prevention: Systematic review and meta-analysis. *Journal of the American Medical Association, 297*(8), 842–857.

Canetti, L., Bachar, E., & Berry, E. M. (2002). Food and emotion. *Behavioural Processes, 60,* 157–164.

Choi, C. (2014, July 9). General Mills' recipe for higher sales: Add fiber, cinnamon. Associated Press. Retrieved from http://minnesota.cbslocal.com

Codex Alimentarious Commission. (1991). General principles for the addition of essential nutrients to foods. (Report No. CAC/GL 09–1987). Retrieved from http://www.codexalimentarius.org

Druckerman, P. (2015, April 23). Eat up. You'll be happier. *The New York Times,* p. A27. Retrieved from http://www.nytimes.com

Eng, M. (2015, March 28). Guess what makes the cut as a "smart snack" in schools? Hot Cheetos [The Salt web log]. National Public Radio. Retrieved from http://www.npr.org

Food additives. (1959, February 27). *The New York Times,* p. 24. Retrieved from http://timesmachine.nytimes.com

Food and Drug Administration [FDA]. (2010, April). Overview of food ingredients, additives and colors. Retrieved from http://www.fda.gov

Food and Drug Administration [FDA]. (2013, August 22). Agency information collection activities; submission for Office of Management and Budget Review; comment request; experimental studies on consumer responses to nutrient content claims on fortified foods. Retrieved October 9, 2013 from: http://federalregister.gov/a/2013-20469

Food fortification in today's world. (2010, January). *Food insight: Current topics in food safety & nutrition*. Retrieved from http://www.foodinsight.org

Freedman, D. H. (2013, July/August). How junk food can end obesity. *The Atlantic*, pp. 68–89.

Kindy, K. (2014, August 17). Food additives on the rise as FDA scrutiny wanes. *Washington Post*. Retrieved from http://www.washingtonpost.com

Miller, A. (2012, February 24). Popular but dangerous: 3 vitamins that can hurt you. *U.S. News & World Report*. Retrieved from http://health.usnews.com

Nestle, M. (2013, August 23). FDA proposes study of consumer attitudes to fortified snack foods [Web log post]. Retrieved from http://www.foodpolitics.com

Noguchi, Y. (2015, March 21). As Americans eat healthier, processed foods start to spoil. National Public Radio. Retrieved from http://www.npr.org

Price, C. (2015, February 15). Vitamins hide the low quality of our food. *The New York Times*, p. SR5. Retrieved from http://www.nytimes.com

Quinn, E. (2015, April 14). Why the FDA has never looked at some of the additives in our food. National Public Radio. Retrieved from http://www.npr.org

Rutsch, P. (2015a, April 15). Not so fast, Kind Bars: FDA smacks snacks on health claims. National Public Radio. Retrieved from http://www.npr.org

Rutsch, P. (2015b, March 6). Voluptuous vet: Can food porn seed lust for healthy eating? [The Salt web log]. National Public Radio. Retrieved from http://www.npr.org

Sacco, J., & Tarasuk, V. (2009). Health Canada's proposed discretionary fortification policy is misaligned with the nutritional needs of Canadians. *The Journal of Nutrition, 139*(10), 1980–1986. doi: 10.3945/jn.109.109637

Sakimura, J. (2013, September 11). Fortification: How food companies fool you. Nutrition sleuth: The truth about food fads, supplements and diet scams [Web log post]. Retrieved from http://www.everydayhealth.com

Schleicher, B. (2007, April 21). Diet Coke Plus. *The Baltimore Sun*. Retrieved from http://articles.baltimoresun.com

Schmeck, H. M. (1974, July 13). F.D.A. asks rules on nutrients added to food products. *The New York Times*, Retrieved from http://timesmachine.nytimes.com

Zimmer, C. (2013, December 9). Vitamins' old, old edge. *The New York Times*, p. D1. Retrieved from http://www.nytimes.com

13
Electronic Cigarettes

In 2014, the word of the year, according to Oxford Dictionaries, was "vape," the colloquial term for inhaling nicotine aerosol from an e-cigarette (Bennett, 2014b). E-cigarettes are battery-powered tubes that heat liquid nicotine, flavorings, and other chemicals into a vapor that can be inhaled (DrugFacts, 2014). Although a patent was given in 1965 for a smokeless, non-tobacco cigarette (Gilbert, 1965), the current version, e-cigarettes, did not come to market until the 2000s, and their usage quickly grew (Rom, Pecorelli, Valacchi, & Reznick, 2015).

Are e-cigarettes a "miracle cure" to smoking or a new vehicle for drug addiction? Do they remove toxins or replace old toxins with new ones? What effects will e-cigarettes have on environmental and societal health? Will e-cigarettes usher in new norms of "cool" identities and ways to socialize? Will they circumvent laws that limit cigarette marketing and use, or become a new focus for regulation? Do they hide drug abuse in plain sight? In our society's attempt to reduce drug dependence, are we leading toward a different flavor of dependence?

Huffing and Puffing about Addiction

with Zachary Goodstein

The "1960s look" may be staging a comeback: hip young folks congregating, raising tubes to their lips at regular intervals. Some tubes actually look like an old-fashioned cigarette complete with a "burning" end, whereas others look like someone is sucking on a writing pen. What's not in the picture? Smoke. Instead, what is exhaled is similar to visible water vapor, like breath on a freezing day.

After the 1964 landmark report that smoking posed health risks (see US Surgeon General, 2014), it took decades for most states to outlaw or limit smoking in restaurants, workplaces, schools, hospitals, theaters, and even some outdoor spaces (List of smoking bans in the United States, 2015). Separately, federal law prohibited smoking in airplanes and federal government buildings (Rutsch, 2015). These "denormalizing" efforts (Clune, 2013) turned a habit linked to cancer into a vice, and they considerably reduced the prevalence of smoking (Esterl, 2013).

Yet, in 2013, almost one in five American adults still smokes cigarettes (Centers for Disease Control and Prevention, 2014). Plus, in 2014, 2.5% of middle school students and 9.2% of high school students reported smoking in the last 30 days (Centers for Disease Control and Prevention, 2015).

A "feel good" way to reduce smoking?

E-cigarettes entered the market as a tool to stop smoking (DrugFacts, 2014). Even health advocates supported them as a pathway to reduce cancer risk (Dreaper, 2014). E-cigarettes don't have the tar considered most responsible for cancer from cigarettes (Meier, 2014), yet they leave intact the pleasurable benefits of nicotine (Etter & Bullen, 2011).

E-cigarette usage is on the rise, mostly by current and former adult smokers (Grana, Benowitz, & Glantz, 2014). One in ten American adults are now "vapers," a four-fold increase in two years, but many vapers also still smoke (Mincer, 2015). Furthermore, since sweet flavorings appeal to young people, youth also have tried the new devices (Grana et al., 2014; Richtel, 2014b). The rate of youth adoption alarms many health advocates (Tavernise, 2014), especially since many of these youth were never smokers (Grana et al., 2014). The worry is that e-cigarettes may be the entry point to nicotine addiction.

E-cigarette manufacturers admit that they haven't yet completely replaced smoking (Nocera, 2014). Plus, although initially against e-cigarettes, the big tobacco companies now back the new devices (Esterl, 2013; Meier, 2014). This shift worries some start-ups that may not have the resources to compete against these large companies (Kershaw, 2014).

Still blowing smoke in the face of regulators?

E-cigarettes have fewer marketing regulations than smoke cigarettes: they can be advertised, and messages can include celebrities, fun social situations, and show e-cigarette usage nearly anywhere without infringing on others' rights to avoid secondhand smoke or local smoke-free policies (Grana et al., 2014; McCullough, 2015).

However, lawmakers around the world have banned or are considering regulation for e-cigarettes (Grana et al., 2014) and their marketing tactics (The Editorial Board, 2015). The World Health Organization increasingly calls for regulations (Mundasad, 2014). US regulators aim to ban marketing to youth, especially as a Surgeon General's report suggests adolescence may be a "critical period" in which nicotine can damage the developing brain (Mickle, 2014).

Truly a breath of fresh air?

The promise of e-cigarettes was the pleasure of smoking without the toxins. But testing found that, when used at high voltage, some e-cigarettes produce formaldehyde (Jensen, Luo, Pankow, Strongin, & Peyton, 2015), also linked to cancer. However, some commentators suggest this finding may be more inflammatory than helpful since most vapers don't use high voltage since it tastes bad, and denigrating e-cigarettes may return vapers to traditional smoking (Nocera, 2015).

A second health concern is the lax oversight of e-cigarette manufacturing (Barboza, 2014). Shoddy production could lead to tiny metal particles making their way into the vapor. A third health concern is that little is known about the long-term consequences of inhaling the "inert" ingredients of the vapor (Bertholon, Becquemin, Annesi-Maesano, & Dautzenberg, 2013).

Perhaps most concerning is the impact of nicotine itself. Mentally, nicotine is a powerful stimulant that affects brain receptors related not only to pleasure – which is why may people smoke – but also to heart rate, blood pressure, appetite, fine motor control, alertness, memory, and mood (National Institute of Drug Abuse, n.d.). On one hand, nicotine reduces symptoms of attention-deficit/hyperactivity disorder

(Levin, Conners, Sparrow, Hinton, et al., 1996) and Parkinson's disease (Kelton, Kahn, Conrath, & Newhouse, 2000). On the other hand, nicotine can be flammable, corrosive, and, when heated, produce oxides and other toxic fumes (National Institute for Occupational Safety and Health, 2014).

In liquid form, nicotine is easily absorbed, and small amounts can cause vomiting, heart problems, seizures, and death – especially for children who may not know what the liquid is and drink it if it tastes sweet (Richtel, 2014a). Little is known about the health effects of interactions among nicotine and the inert chemicals in e-cigarettes, especially since the composition of e-cigarette liquids range widely and may not be properly labeled (Bertholon et al., 2013).

Addiction prediction: "High-ing in plain sight"

Nicotine is as addictive as heroin and cocaine (National Institute of Drug Abuse, n.d.), and it may be a "gateway drug" to marijuana and cocaine use (Kandel & Kandel, 2014). As a result, e-cigarettes as "pure nicotine delivery devices" may prime the brain for further drug use, thus creating health risks unrelated to smoke (Kandel & Kandel, 2014).

Furthermore, young people find the invisibility and harmlessness of e-cigarettes appealing. Young vapers like that they could vape in places where smoking is not allowed, such as at school or work (Peters, Meshac, Lin, Hill, & Abughosh, 2013). Since e-cigarettes are easy to hide and create fewer tell-tale signs – like cigarettes' secondhand smoke or cigarette butts – vaping makes an easy "quick fix." Plus, refillable e-cigarettes can be altered to use with marijuana (Bryan, 2014), so it is possible that vapers could be getting high on a variety of substances without others around them noticing. Although many airlines prohibit e-cigarettes under the federal law banning smoke, would standard smoke detectors pick up the vapor?

Potential longer-term vaper capers

As media focus on the fewer health harms of vaping versus smoking on the vapers' own health, questions about e-cigarettes' effects on the well-being of others and the environment are backgrounded. Like smoking, vaping produces air pollution (Grana et al., 2014). E-cigarettes might be considered *more* dangerous than smoke to people around vapers because, unless the vapor is scented, bystanders may not even realize they are inhaling nicotine aerosol. With exposure over time, non-vapers might find themselves addicted to nicotine.

Furthermore, e-cigarettes produce waste. Like other electronics, e-cigarette tubes contribute metal waste and batteries contribute toxic waste. Liquid nitrogen cartridges contribute traces of a neurotoxin that can seep into the ground or water or be absorbed by trash-scouring insects and animals. Disposable e-cigarettes contribute to the landfill faster.

"Hey, buddy, can I bum a vape?"

Harms to one's own or to others' health are not the only ethical quandaries with e-cigarettes. Vaping presents potential social concerns or opportunities, depending on our perspective. There may be benefits to current smokers who, with e-cigarettes, might enjoy their habit in locations and situations where previously they were shunned due to secondhand smoke and tobacco smells. Whereas smokers in some states were limited to their own homes and cars, now they can be seen vaping in public. Our tendency toward social mimicry may lead to increased vaping just to fit in or to be seen with the "cool crowd" – vaping becomes the norm (Tavernise, 2014).

This scenario becomes increasingly likely as social infrastructure develops that supports a vaping culture. For example, vape shops are popping up in major cities, and some might become social centers where enthusiasts can compare flavor preferences, or become workshops to rebuild parts of the vaporizer, or mimic the vibe of a hookah bar where friends vape together. Perhaps, these venues might hold events, like e-cigarette "tastings." Critics – like movie reviewers – might arise to compare and contrast new flavors (as occurred once marijuana was legalized; Bennett, 2014a). E-cigarettes might spawn a line of accessories, turning them into a fashion focus. Vapers might own several e-cigarettes to match their moods or their outfits. Perhaps, e-cigarette manufacturers might come out with new styles each year to keep demand high.

Over time, as this infrastructure and culture grows, even nonsmokers may join in. Youth, especially, may start to view e-cigarettes as a way to rebel against authority (McCullough, 2015). Although the majority of youth who have tried e-cigarettes so far do not consider them "cool" (Kong, Morean, Cavallo, Camenga, & Krishnan-Sarin, 2015), attitudes toward vaping could become increasingly favorable, just as occurred with marijuana over time. Vaping may return smoking to its pinnacle of coolness – like in the 1960s – but without the smoke.

Further exploration

1. Take the role of a business owner or school principal. What policy do you think would be appropriate to implement regarding vaping at your site?
2. What are some short-term and long-term implications of social interactions that are fundamentally built around an addiction (such as vaping parties or hangouts)?
3. Compare smoking and vaping on ethical impacts beyond personal or public health. What might be the different effects they have on how people treat each other, contribute to society, and/or aim for a better world?

References

Barboza, D. (2014, December 13). China's e-cigarette boom lacks oversight for safety. *The New York Times*. Retrieved from http://www.nytimes.com

Bennett, J. (2014a, November 9). Clean, with a note of citrus. *The New York Times*, p. S1, 8. Retrieved from http://www.nytimes.com

Bennett, J. (2014b, November 23). "Vape" joins pot's lingo, with hat tip to Oxford. *The New York Times*, p. S14. Retrieved from http://www.nytimes.com

Bertholon, J. F., Becquemin, M. H., Annesi-Maesano, I., & Dautzenberg, B. (2013). Electronic cigarettes: A short review. *Respiration*, 86(5), 433–438. doi: 10.1159/000353253

Bryan, M. (2014, April 18). Pot smoke and mirrors: Vaporizer pens hide marijuana use [Shots web log]. National Public Radio. Retrieved from http://www.npr.org

Centers for Disease Control and Prevention. (2014). Current cigarette smoking among adults – United States, 2005–2013. *Morbidity and Mortality Weekly Report 2014*, 63(47), 1108–1112. Retrieved June 10, 2015, from http://www.cdc.gov

Centers for Disease Control and Prevention. (2015). Tobacco use among middle and high school students – United States, 2011–2014. *Morbidity and Mortality Weekly Report 2015*, 64(14), 381–385. Retrieved June 10, 2015, from http://www.cdc.gov

Clune, S. (2013, July 8). The real reason behind public smoking bans. *PBS NewsHour*. Retrieved from http://www.pbs.org

Dreaper, J. (2014, May 29). "Resist urge to control e-cigarettes," WHO told. British Broadcasting Corporation. Retrieved from http://www.bbc.com

DrugFacts. (2014, September). DrugFacts: Electronic cigarettes. National Institute on Drug Abuse. Retrieved June 10, 2015 from http://www.drugabuse.gov

Esterl, M. (2013, November 18). Big Tobacco begins its takeover of the e-cigarette market. *Wall Street Journal*. Retrieved from http://www.wsj.com

Etter, J. F., & Bullen, C. (2011). Electronic cigarette: Users profile, utilization, satisfaction and perceived efficacy. *Addiction*, 106(100), 2017–2028. doi: 10.1111/j.1360-0443.2011.03505.x

Gilbert, H. A. (1965). U.S. Patent No. 3200819 A. Washington, DC: U.S. Patent and Trademark Office. Retrieved from http://www.google.com/patents/US3200819.

Grana, R., Benowitz, N., & Glantz, S. A. (2014). E-cigarettes: A scientific review. *Circulation, 129*, 1972–1986. doi: 10.1161/CIRCULATIONAHA.114.007667

Jensen, R. P., Luo, W., Pankow, J. F., Strongin, R. M., & Peyton, D. H. (2015, January 22). Hidden formaldehyde in e-cigarette aerosols [Letter to Editor]. *New England Journal of Medicine, 372*, 392–394. doi: 10.1056/NEJMc1413069

Kandel, E. R., & Kandel, D. B. (2014). A molecular basis for nicotine as a gateway drug. *The New England Journal of Medicine, 371*, 932–943. doi: 10.1056/NEJMsa1405092

Kelton, M. C., Kahn, H. J., Conrath, C. L., & Newhouse, P. A. (2000). The effects of nicotine on Parkinson's disease. *Brain and Cognition, 43*(1–3), 274–282.

Kershaw, O. (2014, October 14). Don't let Big Tobacco crush e-cigarettes. Cable News Network. Retrieved from http://www.cnn.com/

Kong, G., Morean, M. E., Cavallo, D. A., Camenga, D. R., & Krishnan-Sarin, S. (2015). Reasons for electronic cigarette experimentation and discontinuation among adolescents and young adults. *Nicotine & Tobacco Research, 17*(7), 847–854. doi: 10.1093/ntr/ntu257

Levin, E. D., Conners, C. K., Sparrow, E., Hinton, S. C., Erhardt, D., Meck, W. H., Rose, J. E., & March, J. (1996). Nicotine effects on adults with attention-deficit/hyperactivity disorder. *Psychopharmacology, 123*(1), 55–63.

List of smoking bans in the United States (2015). Wikipedia. Retrieved June 10, 2015 from http://en.wikipedia.org

McCullough, M. (2015, March 9). E-cigs get latitude on ads – which troubles some. *The Philadelphia Inquirer*. Retrieved from http://articles.philly.com

Meier, B. (2014, December 24). Race to deliver nicotine's punch, with less risk. *The New York Times*. Retrieved from http://www.nytimes.com

Mickle, T. (2014, November 13). E-cigarette use by teens rising. *The Wall Street Journal*. Retrieved from http://www.wsj.com

Mincer, J. (2015, June 10). E-cigarette usage surges in past years: Reuters/Ipsos poll. *Reuters*. Retrieved from http://www.reuters.com

Mundasad, S. (2014, August 26). "Ban e-cigarette use indoors," says WHO. British Broadcasting Corporation. Retrieved from http://www.bbc.com

National Institute of Drug Abuse. (n.d.). Parent guide – Legal doesn't mean harmless. Retrieved June 10, 2015, from http://www.drugabuse.gov/sites/default/files/parentguidemod2_69_0.pdf

National Institute for Occupational Safety and Health. (2014, November 20). Nicotine: Systemic agent. Retrieved from http://www.cdc.gov/niosh

Nocera, J. (2014, November 28). Nicotine without death. *The New York Times*. Retrieved from http://www.nytimes.com

Nocera, J. (2015, January 27). Is vaping worse than smoking? *The New York Times*, p. A27. Retrieved from http://www.nytimes.com

Peters, R. J., Meshack, A., Lin, M.-T., Hill, M., & Abughosh, S. (2013). The social norms and beliefs of teenage male electronic cigarette use. *Journal of Ethnicity in Substance Abuse, 12*(4), 300–307. doi:10.1080/15332640.2013.819310

Richtel, M. (2014a, March 24). Selling a poison by the barrel: Liquid nicotine for e-cigarettes. *The New York Times*, p. A1. Retrieved from http://www.nytimes.com

Richtel, M. (2014b, July 15). E-cigarette makers are in an arms race for exotic vapor flavors. *The New York Times*, p. A1. Retrieved from http://www.nytimes.com

Rom, O., Pecorelli, A., Valacchi, G., & Reznick, A. Z. (2015, March). Are e-cigarettes a safe and good alternative to cigarette smoking? *Annals of the New York Academy of Sciences, 1340*, 65–74. doi: 10.1111/nyas.12609

Rutsch, P. (2015, February 24). Will vaping reignite the battle over smoking on airplanes? [Shots web log]. National Public Radio. http://www.npr.org

Tavernise, S. (2014, December 16). E-cigarettes top smoking among youths, study says. *The New York Times*. Retrieved from http://www.nytimes.com

The Editorial Board. (2015, April 23). The perils of smokeless tobacco [Editorial]. *The New York Times*, p. A26. Retrieved from http://www.nytimes.com

US Surgeon General. (2014). *The health consequences of smoking – 50 years of progress*. Washington, DC: US Department of Health and Human Services. Retrieved from http://www.surgeongeneral.gov

Part IV
Emotion Tuners

14
Manipulating Emotions

Commercial drugs designed to induce or manipulate emotions are not new. Caffeine and alcohol have been used for centuries to perk us up, calm us down, and smooth social encounters (Young, 2003). More recent findings that acetaminophen may reduce emotional pain (Bakalar, 2015) might lower our perception of social harms, and discussion of a popular wrinkle-removal procedure's possible influence on our emotional expressiveness (Mundasad, 2014) might lead to stunted social development.

In the 1980s, the first selective serotonin re-uptake inhibitor (SSRI) antidepressant became not only a medical treatment but a cultural game-changer. It changed the way we thought about mood disorders. They were not character flaws or overwrought responses to bad situations. They were chemical imbalances in the brain (Haberman, 2014).

With that shift in meaning, the gold rush was on to develop "happy pills" and other pharmaceutical means to remove personality, behavior, or lifestyle nuisances. Let's bottle "happily ever after"! These chemicals might even out the rollercoaster of life even if no disease is present. Another way to consider the situation: life problems could be medicalized so they would be eligible for drug interventions (Hardy, 2012). If we don't feel just right, see our doctors and the problem might be easily solved (Lexchin, 2001).

The innovation of intentionally manipulating emotions with chemicals is twofold. First, emotions become no longer visceral indicators of our interactions with the environment: contentment is not a signal of person-environment alignment, frustration does not tell us our goal is blocked and we need to find another way, and disgust or fear does not indicate something to avoid. Rather, emotions become a disruption of a smooth-sailing life: they interrupt our productivity, mindless entertainment, and self-conceptions that we've "got it all under control."

Second, the chemicals under consideration for market exploitation are hormones and nutrients already found in the body for which researchers are discovering potential uses. It may be difficult to determine a line between genuine and chemically induced emotions (similar to performance-enhancement drugs), and there may be considerable unintended consequences since many hormones perform several functions.
What ethical side-effects might we face? Will we feel "groovy" because we never get down on ourselves or others? Or will we become cruel because we don't have the emotional feedback loop to tell us we've hurt others? What becomes of our "self" if our feelings have flatlined into a vague contentment?

Feelings for Sale
with Curtis Meyer

To an extent, all of life involves the manipulation of chemicals. But upcoming innovations in biochemistry produce new opportunities to manipulate feelings – possibly without us even being aware of it. The chemicals most in the news these days are natural substances in our bodies already, but researchers have figured out new ways they may be useful: nutrients like Vitamin C and omega-3 oils to reduce the emotional effects of stress (Young, 2003); neurotransmitters like dopamine and serotonin that boost our mood (Haberman, 2014; Sweeney, 2005); and hormones like testosterone and oxytocin that influence social emotions and behaviors like negotiation, aggression, lust, empathy, and trust (Eisenegger, Naef, Snozzi, Heinrichs, & Fehr, 2010; Kosfeld, Heinrichs, Zak, Fischbacker, & Fehr, 2005; Sweeney, 2005).

How might this innovation affect who we are as persons, our interpersonal interactions, and our social institutions? Do we care what generates our emotions? Do we care who controls that generation? What is *genuine* emotion? Emotions affect perception, decision-making, and social relationships. They tell us what to approach and what to avoid. They can put us in or keep us out of dangerous situations. They contribute to reasoning (Damasio, 2005), but are often depicted as irrational interference in problem-solving. Should they be celebrated or eradicated? If we proceed with developing chemicals to manipulate emotions before we are ready to handle the ramifications, are we risking what it means to be human?

Manipulating ourselves

Emotions are part of our sense of self. They comprise a large part of our subjective experience. If I can use a pill or nasal spray to make myself feel a particular emotion on demand, who am "I"? While emotions emerge from chemical releases in the brain, we experience emotion as resulting from life events. We feel proud when we score the winning point. We feel angry when insulted. With no event, we have no anchor to make sense of the bodily sensation (Schachter & Singer, 1962). If we discount emotions because they come conveniently from a pill, it may devalue the meaning of our life. There is nothing to care about because we conceive caring – emotional involvement – as irrelevant. It's just a chemical.

This situation may segue into loss of motivation. We lose the elation of a job well done, or the impetus to do better if our performance is under par, or the courage to meet new friends. Why bother? Actions as emotional causes may become misleading when any emotion can be conjured easily with a pill. However, chemically manipulated emotions may lead to actions we might not engage in otherwise. Testosterone may make us more lustful and aggressive in getting what we want, even if we are usually considered "calm, cool, and collected."

This unfortunate situation may arise particularly since we may not know the proper dose for the desired outcome. For example, oxytocin invokes feelings of tenderness and empathy toward others (Angier, 2009), but it also may increase in-group favoritism (Wade, 2011). So if one peace negotiator, say, takes a lower dose and becomes more trusting, but the other negotiator takes more and considers only his own group, an agreement may be reached, but it may be unfair.

Furthermore, chemical manipulation of emotions may have unintended consequences because both the chemicals and the emotions involve complex processes. For example, the neurotransmitter dopamine not only contributes to romance (Sweeney, 2005) through its "reward circuits," but also affects movement, concentration, and memory. Antidepressants that affect serotonin, another neurotransmitter, make people feel less sad but also can increase sleepiness and fatigue. Indiscriminate use of these chemicals as lifestyle enhancers could have far-reaching effects on our well-being – not all of which we would want. We might get caught up in a cycle of using one chemical to offset another chemical's effects. Managing our emotions could be a pharmacological Gordian knot that we can't undo.

Similarly, emotions involve several dimensions: appraisal of a situation, such as "that person cut in line"; the visceral response of a churning stomach and flushed skin; cognitive recognition that "I am angry"; and expression through a scowl, a shout, and perhaps a shove to move the person away (Scherer, 2005). Chemically manipulated emotions eliminate the role of an event, so we are left with unanchored bodily sensations and actions.

Emotional reactions to events can settle over time into moods or understandings about oneself: too many negative experiences may induce a low self-worth or too many positive experiences a sense of invincibility. Thus, emotions help define who we become (Weiss & Beal, 2005). If we chemically manipulate our emotions to always be pleasant, we miss out on the character-building that comes from facing hardships and challenges (Shulevitz, 2015) and increase our fragility and vulnerability in a vicious cycle (Iarovici, 2014). Even if we add chemicals that might induce the feelings of frustration or confusion, our character may become nothing more than a long chemical formula of what we've ingested. Almost all of personality psychology might become useless when extroversion and neuroticism might become types of pills, not types of people.

Manipulating each other

We are a techno-focused society, chasing external gadgets and tools – like mediated communication apps and emoticons – to fill the emotional gap between ourselves and others. These tools reduce the messiness of immediate social interaction because we can send short one-sided messages via the Internet, then read and respond at our convenience. With our focus on technology, non-techie options to enhance positivity in social situations may sound old-fashioned: hugs and kisses (Kabilan, 2014), honesty, or genuine praise toward or interest in another person (Hardy, 2012).

With chemical manipulation of emotions, we may not need intermediary gadgets. We change ourselves or each other into more socio-emotionally adept people. The current "media sensation" hormone is oxytocin, nicknamed "the love hormone" (Reynolds, 2012). Studies find it helps build empathy, trust, and attachment (Schaller, 2007; Zak, Kurzban, & Matzner, 2005) – perhaps excessively so, since it may induce us to trust and attach when we perhaps shouldn't. For example, a spritz of oxytocin through the nose makes people more likely to trust a stranger with their money (Angier, 2009). An oxytocin body spray that others could inhale (Glassie, 2005) could put us at an advantage on dates, or as a salesperson, stockbroker, financial adviser, realtor, or other persuasion-oriented position. Would this manipulation be fair, since we are likely to misattribute the reason for our amenability to the person's skill and not a hormone? Oxytocin also seems helpful for building team spirit (Reynolds, 2012), so it could aid situations with sports (Reynolds, 2012), the military (Wade, 2011), cause-based activist groups, and collaborative businesses (Hardy, 2012). Why go through the messy stages of authentic team formation when a whiff of oxytocin makes everyone play well together? However, oxytocin also may exacerbate teams of diverse individuals, political negotiations, and creative teams. Finally, oxytocin may stimulate antisocial behaviors, such as gloating and envy (Angier, 2009). So it counterproductively might confuse social relations as a person at one moment is hugging us and the next moment is acting condescendingly.

Is an oxytocin spray "cheating" at sociability, social graces, and social norms? We would not need to cope with the death of a loved one, loss of a job, or disappointment of a betrayal. Perhaps finding a life partner entails simply slipping oxytocin into a meal. How will we know our new love is real love? Perhaps, interpersonal skills would no longer be needed because everybody would be able to chemically obtain social graces. Oxytocin sprays seem more convenient than directly engaging the social situation honestly and having to go through the work of making a true friend. But how would we distinguish who are our friends or acquaintances, lovers or foes, since oxytocin boosts our trust in everyone?

These chemicals might become crutches. A tool to help us become more socially adept ends up making us socially inept without them. Family dynamics might be great until the oxytocin in the air conditioning system runs out. Then what? Siblings may no longer have the wherewithal to overcome rivalries. The love triangles common in Shakespearean plays might be more likely to end as tragedies since the chemically induced social connections leave us ill-equipped to deal with rejection. There is no guarantee that the oxytocin entices our beloved to love us – their sights might be set on someone else. To get over it, we might need to counter with a different chemical – testosterone if we want revenge, an antidepressant if we want to blunt the pain.

Yet, if we never feel emotional pain, will other people become viewed as simply resources for us? Will we treat others with respect because we know what it's like to feel disrespected? Or will we become bullies – especially if our testosterone is high – because, although our oxytocin makes us trust others, it doesn't mean they will fulfill our needs. If they don't, without chemical inducement, will we care for them or disregard them? Could we become cruel in the pursuit of our desires? Chemicals that make us fearless – such as those used by soldiers in war (Lin, 2012; Plaue, 2012) – could lead to social risk-taking. Those risks may not work out as we'd like. The fearlessness may inspire a relentless pursuit of what we want – perhaps even more ruthlessly because the oxytocin made us trust the other person and we feel betrayed.

Policies for emotional reform?

If chemical manipulation of emotion makes hospital patients calmer and more trusting of providers, soldiers more focused and fearless, students and workers more productive team players, why not institutionalize the use of these chemicals? What if their use became no longer voluntary and self-selecting, but required, say, for a particular career or to receive healthcare or air travel? Instead of selecting employees based on character traits, create a chemical regimen to produce the character desired.

This scenario might make it easier for employers to keep their companies humming along, airlines to maintain order in economy class, and cancer patients to sustain hope. It might reduce the emotional variability in a situation without the need for leadership or self-regulation. The hope is that these situations would proceed without emotional incident – everyone is calm and *un*emotional. But this scenario also might create a moral hazard because, without the emotional feedback indicating when someone has "had enough," the situation might escalate. Companies might increase work hours, doctors up chemotherapy dosages, and the like, until people die with no warning. Their emotions did not give signals of how they were doing.

Perhaps these dire incidents could lead to mandated emotional reform. But who should have the power to decide who does and who doesn't need a chemically induced "emotional adjustment"? Could we lose our jobs because we weren't suitably happy one day? Could schools make it a requirement, for safety reasons, to take hormones to avoid expression of anything but contentment? Would there be informed consent – or any knowledge at all – by shoppers that a store blew oxytocin through its ventilation system to stimulate us to trust the sales clerk and buy the up-sell products?

What if a rebellion ensues against chemical manipulation of emotions? "Loyalists" who stay on oxytocin continue to feel moderately ok. But those who abstain and protest might be at a disadvantage because dependence on the chemicals makes feeling anything confusing. They may become overwhelmed by all the "unwanted" emotions that the chemicals kept at bay (Iarovici, 2014). Or they may lack motivation to proceed with the protest. The rebellion may not last long, since the activists may no longer have the skills to make sense of and to regulate their emotions.

Further exploration

1. What innovative techniques might we develop to decipher whether our own feelings or others' emotional expressions are genuine (not chemically induced)?
2. How might our notion of "relationship" change as emotions become so easy to chemically induce in ourselves and others?
3. Pick a type of relationship – friendship, dating, business partnership, doctor-patient, etc. Assume that all emotions are now chemically induced. Write a brief story that shows what a social interaction of this type of relationship might look and sound like.

References

Angier, N. (2009, November 23). The biology behind the milk of human kindness. *The New York Times,* p. D2. Retrieved from https://www.nytimes.com

Bakalar, N. (2015, April 20). Tylenol may blunt emotions, and not just pain [Well web log]. *The New York Times.* Retrieved from https://www.nytimes.com

Damasio, A. (2005). *Descartes' error: Emotion, reason, and the human brain* (Rev. ed.). New York, NY: Penguin.

Eisenegger, C., Naef, M., Snozzi, R., Heinrichs, M., & Fehr, E. (2010, January 21). Prejudice and truth about the effect of testosterone on human bargaining behaviour. *Nature, 463,* 356–359.

Glassie, J. (2005, December 11). Trust spray. *The New York Times.* Retrieved from https://www.nytimes.com

Haberman, C. (2014, September 21). Selling Prozac as the life-enhancing cure for mental woes. *The New York Times.* Retrieved from https://www.nytimes.com

Hardy, Q. (2012, June 25). Will oxytocin keep the virtual office humming? [Web log post]. *The New York Times.* Retrieved from http://bits.blogs.nytimes.com

Iarovici, D. (2014, April 17). The antidepressant generation [Web log post]. *The New York Times.* Retrieved from http://well.blogs.nytimes.com

Kabilan, A. (2014). Pharmacological role of oxytocin – a short review. *Journal of Pharmaceutical Science and Research, 6*(4), 220–223.

Kosfeld, M., Heinrichs, M., Zak, P. J., Fischbacker, U., & Fehr, E. (2005, June 2). Oxytocin increases trust in humans. *Nature, 435*, 673–676.

Lexchin, J. (2001). Lifestyle drugs: Issues for debate. *Canadian Medical Association Journal, 164*, 1449–1451.

Lin, P. (2012, February 16). More than human? The ethics of biologically enhancing soldiers. *The Atlantic*. Retrieved from http://www.theatlantic.com

Mundasad, S. (2014, September 13). Botox "may stunt emotional growth" in young people. British Broadcasting Corporation. Retrieved from https://www.bbc.com

Plaue, N. (2012, June 15). There's a drug that might get rid of all our fears. *Business Insider*. Retrieved from http://www.businessinsider.com

Reynolds, G. (2012, November 21). The "love hormone" as sports enhancer [Web log post]. *The New York Times*. Retrieved from http://well.blogs.nytimes.com

Schachter, S., & Singer, J. (1962). Cognitive, social, and physiological determinants of emotional state. *Psychological Review, 69*, 379–399. doi:10.1037/h0046234

Schaller, M. (2007). Is secure attachment the antidote to everything that ails us? *Psychological Inquiry, 18*(3), 191–193.

Scherer, K. R., (2005). What are emotions? And how can they be measured? *Social Science Information, 44*, 693–727.

Shulevitz, J. (2015, March 21). In college and hiding from scary ideas. *The New York Times*. Retrieved from http://www.nytimes.com

Sweeney, C. (2005, June 5). Not tonight. *The New York Times*. Retrieved from https://www.nytimes.com

Wade, N. (2011, January 10). Depth of the kindness hormone appears to know some bounds. *The New York Times*, p. D1. Retrieved from https://www.nytimes.com

Weiss, H. M., & Beal, D. J. (2005). Reflections on affective events theory. *Research on Emotion in Organizations, 1*, 1–21. doi: 10.1016/S1746-9791(05)01101-6

Young, S. N. (2003). Lifestyle drugs, mood, behaviour and cognition. *Journal of Psychiatry & Neuroscience, 28*(2), 87–89.

Zak, P. J., Kurzban, R., & Matzner, W. T. (2005). Oxytocin is associated with human trustworthiness. *Hormones and Behavior, 48*, 522–527.

15
Legalized Marijuana

"Cannabusiness ... cannasseur ... budtender" (Bennett, 2014). These are the buzzwords for the budding industry of legal recreational marijuana. These businesses are regulated and face hurdles – like getting legal bank accounts (Gittleson, 2014). But the industry is, well, growing like a weed – sometimes in unexpected ways: vending machines (Chappell, 2014), food trucks (Runyon, 2014b), chefs and foodies experimenting with recipes (Severson, 2014), B&Bs that serve "bud and breakfast" (Bennett, 2014), wedding planners who "toke" their client's big day to a new "high" (Brady, 2014). Investors have taken notice (Miller, 2014; Sorkin, 2015). Hopeful entrepreneurs from across the nation are moving in on the action (Healy & Johnson, 2014).

Marijuana has moved beyond "zero tolerance" laws that, some say, clogged prisons with nonviolent criminals (Holloran, 2014); beyond counterculture's smoke, giggles, and munchies in a camper van; beyond the "medical miracle" of alleviating pain when nothing else would. Marijuana has gone upscale with bright stores and a variety of options (Runyon, 2014a). Will this high last, or are we headed for a crash?

The Highs and Lows
with Achu Johnson Alexander

Ups and downs in the law

Marijuana's relationship with US law has gone through ups and downs since the early 20th century as political and social sentiments shifted from fear and discrimination, to tough penalties, to leniency, to a War on Drugs (Mannes, 1998). Yet, it has remained a popular recreational drug in the US for decades (Office of National Drug Policy Control, 2010). Although recreational marijuana remains a federal crime (Keefe, 2013), on January 1, 2014, Colorado was the first state to legalize it for adults (Ferner, 2013). Several other states are following suit (Knox, 2014). For the first time in 40 years, more than half of Americans support legalization (Pew Research Center, 2013), and pot use is increasing (Neuman, 2014).

Does "legal" equal "good"? This shift has worldwide repercussions (BBC Staff, 2014b; Johnson, 2014a). Is legal weed a "socially responsible" investment or an evolving "sin industry" (Sorkin, 2015)? Are we correcting an injustice (Wegman, 2014), opening the gates to irresponsibility (Neighmond, 2014), or simply brewing confusion (Parker-Pope, 2014)? Not only is health or business affected, but also psychological, social, and societal expectations.

The highs

Legalized marijuana may launch a renaissance in open-mindedness. Without fear of prosecution, marijuana's effects on perception could be helpful for creative and entrepreneurial opportunities (Healy & Johnson, 2014). Marijuana and other drug use have been associated with creative people for decades (e.g., Tolson & Cuyjet, 2007), although its efficacy as a creativity enhancer is debated (Bourassa & Vaugeois, 2001; Grossman, Goldstein, & Eisenman, 1974). But it does seem to make people at least *feel* like they are more open to new experience, especially if they believe that those effects are what is *supposed* to happen with marijuana (Hicks, Pedersen, Friedman, & McCarthy, 2011). This creativity parallels the cultural shifts borne of the Millennial generation, who consider freedom a given not a goal, and life as an unlimited journey of self-exploration (Arnett, 2014).

Sales from legalized marijuana can fund the common good. Marijuana sales are taxed (Ferner, 2013). Although primarily earmarked for substance abuse treatment and education for youth (Singh, 2014), marijuana taxes are expected to be a "bonanza" for state governments (Healy, 2014b), perhaps contributing to improved safety, schools, or infrastructure. The large sums are creating tension between states and cities, who want a bigger share since local police are most likely to deal with car accidents or overdoses (Koerth-Baker, 2014; Wilson, 2015).

Legalization may alleviate injustices from the War on Drugs. The US has a large prison population (Bureau of Justice Statistics, 2011), in part due to non-violent marijuana convictions (Federal Bureau of Investigation, 2011). Stiff sentences for small amounts of marijuana ruined many people's lives (Wegman, 2014). Some ethnic and socioeconomic groups that have been linked with marijuana (Mannes, 1998; Special Correspondence, 1934) exacerbated differential arrest treatments (Wegman, 2014). Activists call for reconsideration of past convictions (BBC Staff, 2014a), and the federal government is considering new rules (Holloran, 2014).

Legalization may (eventually) improve public safety. Drug enforcement costs are high and accelerating (Sayegh, 2013). Some political leaders justify decriminalization on the idea that, without arrest quotas and court appearances for marijuana offenses, police can concentrate on more serious threats. However, others worry about the short-term problems when legal distributors and street drug dealers compete for the growing market of users (Keefe, 2013). This transition period may require *more* policing to enforce the new law, which potentially might increase dealer arrests or complaints of police harassment. Even legal medical marijuana providers, which previously enjoyed more leniency than recreational marijuana use, may be antagonistic (Johnson, 2014c). Yet, if this transition period is successfully navigated, and new norms are adhered to, marijuana may become a staple in social life.

The lows

"Weed-onomics" is fraught with ethical uncertainties. Many banks refuse to open accounts for pot businesses for fear of federal racketeering charges (Gittleson, 2014). In 2014, federal authorities issued banking guidelines (Kovaleski, 2014), and a few small community banks opened accounts. But, there is an unnerving uncertainty that this legal leniency may be revoked without warning as new politicians come into power. If that occurs, many lives and livelihoods could be irrevocably damaged. Prison populations could swell as bankers, food manufacturers, and others might be caught in a "dragnet" as accomplices. Yet, without bank

accounts, marijuana businesses must deal in cash, which creates risks of robbery and violence (Gittleson, 2014).

Unclear product labeling may harm consumers, especially as the diversity of weed-infused products grows. Food products have been the most problematic to date. Pot could be in snacks to street food to fine dining (Runyon, 2014c; Severson, 2014). Perhaps, over time, marijuana may find its way into perfumes or cosmetics that can be absorbed through breathing or the skin. Public awareness campaigns are a key component of most legalization laws (Singh, 2014), but if labels are unclear about whether, what type, how much, and the quality of marijuana is in a product (Krivonen, 2014; Rutsch, 2015), people might ingest it unwittingly or in too high a dose. The repercussions could be hallucinations and bizarre experiences (Dowd, 2014; Healy, 2014c) or death (Healy, 2014c; Nicholson, 2014).

The proliferation of marijuana usage could breed mistrust among neighbors, workers, and community members. Soon after legalization, next-door neighbors (Gurman, 2014), homeowner associations (Associated Press, 2014), workplaces (Noguchi, 2014), cities (Johnson, 2014b), and surrounding states (Healy, 2014a, 2014c) complained about the corollary effects of widespread marijuana. The highs are felt individually and locally, but the lows can ripple widely through exploding homes (Gurman, 2014), impaired judgment and productivity (Neighmond, 2014), driving under the influence (Koerth-Baker, 2014), and pot crossing borders into states where it's not legal (Healy, 2014c).

Companies in states where pot is legal may be hurt in their bids for federal contracts if the federal government is uncertain that these businesses can or will enforce the federal Drug-Free Workplace Act (Noguchi, 2014). The psychedelic properties of marijuana may make employers question employees' reliability, workers question co-workers' safety, and consumers question companies' service. Will we trust others' professionalism, judgment, confidentiality – especially in sensitive or high-risk work fields? On the other hand, marijuana use is not illegal, so advocates argue for loosening workplace "zero tolerance" policies and on-the-job drug testing as a matter of individual legal rights (Editorial Board, 2014). However, the Colorado Supreme Court decided being under the influence while on the job is cause for dismissal (Peralta, 2015).

Legalized marijuana may make it easier for people to manipulate each other. Marijuana can increase sociability, relaxation, and appreciation of others (Koob, Arends, & Le Moal, 2014). These qualities seem like positives, but marijuana might be used intentionally to manipulate people without their knowing. Possibilities include marijuana-laced products served at business functions to smooth negotiations, at political rallies to promote good feelings toward a candidate, or on flights so passengers stay calm. Unintended consumption of pot, especially with the more potent versions now available, can have dire consequences immediately (Dowd, 2014), and its long-term effects are not well understood (Knox, 2014).

Legalizing marijuana sends mixed messages to future generations. All of the above highs and lows of marijuana legalization become more acute when considering children, teens, and emerging adults. Young people are particularly susceptible to the immediate highs of using marijuana since they don't know as much about the drug. And research suggests that marijuana use during adolescence may interfere with brain development, especially areas important for judgment and problem-solving (Neighmond, 2014), which could put teen users at a disadvantage intellectually, socially, and economically for the rest of their lives. Furthermore, marijuana sales, purchase, or use by people under age 18 remain illegal even in states where they are legal for adults. So youth can still be arrested and end up with a criminal record, which may continue to exacerbate racial discrimination (Gonzalez, 2015).

Yet, pot use among youth is on the rise (O'Connor, 2013). Addiction experts are worried that legalization makes marijuana not only easier to acquire, but also more desirable, as adults more openly promote and role model its use (Parker-Pope, 2014). The types of products that contain marijuana – such as sweets – are strongly desired by children, and children are less likely to critically read package labels (Healy, 2014d). Given the mixed messages between individuals they know and public education campaigns, how are children to figure out what is right?

What may be particularly troubling is a vision of the future in which marijuana legalization ripples into a social norm eschewing social responsibility. Are we ushering the next generation of emerging adults – the Millennials – into a mindset of unlimited individual freedom and possibilities, devoid of relational responsibilities? Early adulthood is a time of exploration and identity development, often coinciding with drug and lifestyle experimentation (Arnett, 2005). Millennials support marijuana more than previous generations (Blow, 2013), and its use is increasing among college students (Moore, 2014). Are we setting up a future society in which few people will have the skills or inclination to address collective problems? Are we paving pathways to "tune out"?

Further exploration

1. What are the pros and cons of governments profiting from a psychoactive drug?
2. Marijuana helps people escape from reality by altering perception and inducing pleasurable feelings not anchored to the current situation. How might marijuana users manage the juxtaposition of "high times" and "real times"? How might this juxtaposition create psychological and social problems?
3. Legalization of marijuana counters one individual's choice to use it with another individual's rights to smoke-free, safe shared spaces. Discuss the tensions involved.

References

Arnett, J. J. (2005). The developmental context of substance use in emerging adulthood. *Journal of Drug Issues, 35*(2), 235–254.

Arnett, J. J. (2014). *Emerging adulthood: The winding road from the late teens through the twenties, 2nd ed.* New York, NY: Oxford University Press.

Associated Press. (2014, July 25). Pot may be legal, but homeowner agreements can ban. Retrieved from http://www.usatoday.com

BBC Staff. (2014a, July 26). The man serving life for marijuana [#BBC trending web log post]. British Broadcasting Corporation. Retrieved from http://www.bbc.com

BBC Staff. (2014b, March 4). UN body criticises US states' cannabis legalisation. British Broadcasting Corporation. Retrieved from https://www.bbc.com

Bennett, J. (2014, October 3). In Colorado, a rebranding of pot inc. *The New York Times*, p. ST1. Retrieved from https://www.nytimes.com

Blow, C. M. (2013, April 5). The young are the restless. *The New York Times*, p. A17. Retrieved from https://www.nytimes.com

Bourassa, M., & Vaugeois, P. (2001). Effects of marijuana use on divergent thinking. *Creativity Research Journal, 13*(3–4), 411–416. doi: 10.1207/S15326934CRJ1334_18

Brady, L. S. (2014, July 27). A toast? How about a toke? *The New York Times*, p. S1.

Bureau of Justice Statistics. (2011). *Justice expenditures and employment, FY 1982–2007 statistical tables* (NCJ 236218). Retrieved from http://www.bjs.gov

Chappell, B. (2014, April 13). Marijuana vending machine unveiled in Colorado [The Two-Way web log post]. National Public Radio. Retrieved from https://ww.npr.org

Dowd, M. (2014, June 3). Don't harsh our mellow, dude. *The New York Times*, p. A23. Retrieved from https://www.nytimes.com

Editorial Board (2014, September 8). Obsolete zero tolerance on pot. *The New York Times*, p. A28. Retrieved from https://www.nytimes.com

Federal Bureau of Investigation (2011). *Uniform crime report – Crime in the United States 2011.* Retrieved from http://www.fbi.gov

Ferner, M. (2013, May 28). Marijuana legalization: Colo. Gov. Hickenlooper signs first bills in history to establish a legal, regulated pot market for adults. *The Huffington Post*. Retrieved from http://www.huffingtonpost.com

Gittleson, K. (2014, February 20). Colorado's marijuana firms beg banks to take their cash. British Broadcasting Corporation. Retrieved from https://www.bbc.com

Gonzalez, D. (2015, April 12). Smoking marijuana for 50 years, and turning out just fine. *The New York Times*, p. A16. Retrieved from https://www.nytimes.com

Grossman, J. C., Goldstein, R., & Eisenman, R. (1974). Undergraduate marijuana and drug use as related to openness to experience. *Psychiatric Quarterly, 48*(1), 86–92.

Gurman, S. (2014, May 6). Hash oil explosions rise with legalized marijuana. *The Denver Post*. Retrieved from http://www.denverpost.com

Healy, J. (2014a, December 18). Nebraska and Oklahoma sue Colorado over marijuana law. *The New York Times*, p. A21. Retrieved from https://www.nytimes.com

Healy, J. (2014b, February 21). Colorado expects to reap tax bonanza from legal marijuana sales. *The New York Times*, p. A12. Retrieved from https://www.nytimes.com

Healy, J. (2014c, May 31). After 5 months of sales, Colorado sees the downside of a legal high. *The New York Times*, p. A14. Retrieved from https://www.nytimes.com

Healy, J. (2014d, October 29). New scrutiny on sweets with ascent of marijuana in Colorado. *The New York Times*, p. A13. Retrieved from https://www.nytimes.com

Healy, J., & Johnson, K. (2014, July 18). Next gold rush: Legal marijuana feeds entrepreneurs' dreams. *The New York Times*, p. A1. Retrieved from https://www.nytimes.com

Hicks, J. A., Pedersen, S. L., Friedman, R. S., & McCarthy, D. M. (2011). Expecting innovation: Psychoactive drug primes and the generation of creative solutions. *Experimental and Clinical Psychopharmacology, 19*(4), 314–320.

Holloran, L. (2014, April 21). Obama seeks wider authority to release drug offenders [It's All Politics web log post]. National Public Radio. Retrieved from https://ww.npr.org

Johnson, G. (2014a, February 15). US drug policy fuels push for legal pot worldwide. Associated Press. Retrieved from http://news.yahoo.com

Johnson, K. (2014b, January 27). Cannabis legal, localities begin to just say no. *The New York Times*, p. A1. Retrieved from https://www.nytimes.com

Johnson, K. (2014c, March 6). Providers on medical marijuana face new fears. *The New York Times*, p. A11. Retrieved from https://www.nytimes.com

Keefe, P. R. (2013, November 18). Buzzkill. *The New Yorker*. Retrieved from http://www.newyorker.com

Knox, R. (2014, March 3). Evidence on marijuana's health effects is hazy at best [Shots web log post]. National Public Radio. Retrieved from https://ww.npr.org

Koerth-Baker, M. (2014, February 18). Driving under the influence, of marijuana. *The New York Times*, p. D1. Retrieved from http://www.nytimes.com

Koob, G. F., Arends, M. A., & Le Moal, M. (2014). *Drugs, addiction, and the brain*. Waltham, MA: Academic Press.

Kovaleski, S. F. (2014, February 15). U.S. issues marijuana guidelines for banks. *The New York Times*, p. A10. Retrieved from http://www.nytimes.com

Krivonen, M. (2014, December 26). Getting high safely: Aspen launches marijuana education campaign. National Public Radio. Retrieved from https://www.npr.org

Mannes, E. (Writer & Producer). (1998). Busted – America's war on marijuana [Television series episode]. In *Frontline*. Retrieved from http://www.pbs.org

Miller, J. (2014, April 4). The misty world of marijuana stocks. British Broadcasting Corporation. Retrieved from https://www.bbc.com

Moore, A. S. (2014, October 29). Legally high at a Colorado campus. *The New York Times*, p. ED14. Retrieved from https://www.nytimes.com

Neighmond, P. (2014, March 3). Marijuana may hurt the developing teen brain [Shots web log post]. National Public Radio. Retrieved from https://ww.npr.org

Neuman, S. (2014, June 26). Pot use on the rise in U.S., report says [The Two-Way web log post]. National Public Radio. Retrieved from https://www.npr.com

Nicholson, K. (2014, April 17). Man who plunged from Denver balcony ate 6x recommended amount of pot cookie. *The Denver Post*. Retrieved from http://www.denverpost.com

Noguchi, Y. (2014, August 12). Colorado case puts workplace drug policies to the test [Shots web log post]. National Public Radio. Retrieved from https://www.npr.com

O'Connor, A. (2013, December 18). Increasing marijuana use in school is reported [Web log post]. *The New York Times*. Retrieved from http://well.blogs.nytimes.com

Office of National Drug Policy Control. (2010). *Fact sheet – Marijuana legalization*. Retrieved from http://www.whitehouse.gov

Parker-Pope, T. (2014, August 18). Legal marijuana for parents, but not their kids [Web log post]. *The New York Times*. Retrieved from http://well.blogs.nytimes.com

Peralta, E. (2015, June 15). Colorado Supreme Court says employees can be fired for marijuana use [The Two-Way web log post]. National Public Radio. Retrieved from https://ww.npr.org

Pew Research Center (2013, April 4). Majority now supports legalizing marijuana. Retrieved from http://www.people-press.org

Runyon, L. (2014a, April 21). To keep business growing, vendors rebrand pot's stoner image. National Public Radio. Retrieved from https://ww.npr.org

Runyon, L. (2014b, August 26). Colorado's pot brownies now come with instructions [The Salt web log post]. National Public Radio. Retrieved from https://www.npr.com

Runyon, L. (2014c, May 14). The latest food truck theme is marijuana for lunch [The Salt web log post]. National Public Radio. Retrieved from https://ww.npr.org

Rutsch, P. (2015, March 24). Quality-testing legal marijuana: Strong but not always clean. National Public Radio. Retrieved from https://www.npr.com

Sayegh, G. (2013, March 20). Marijuana arrests in NYC cost one million police hours. *The Huffington Post*. Retrieved from http://www.huffingtonpost.com

Severson, K. (2014, December 28). Pot pie, redefined? Chefs start to experiment with cannabis. *The New York Times*, p. A1. Retrieved from https://www.nytimes.com

Singh, M. (2014, September 17). Colorado tries hard to convince teens that pot is bad for you [Shots web log post]. National Public Radio. Retrieved from https://www.npr.com

Sorkin, A. R. (2015, January 12). Ethical questions of investing in pot. *The New York Times*, p. B1. Retrieved from https://dealbook.nytimes.com

Special Correspondence. (1934, September 16). Use of marijuana spreading in west; poisonous weed is being sold quite freely in pool halls and beer gardens. Children said to buy it. Narcotic Bureau officials say law gives no authority to stop traffic. *The New York Times*, p. 72. Retrieved from https://timesmachine.nytimes.com

Tolson, G. H., & Cuyjet, M. T. (2007). Jazz and substance abuse: A road to creative genius or a pathway to premature death. *International Journal of Law and Psychiatry 30*, 530–538.

Wegman, J. (2014, July 28). The injustice of marijuana arrests. *The New York Times*, p. A20. Retrieved from https://www.nytimes.com

Wilson, C. (2015, February 8). Cities argue for a bigger share of pot tax revenue. National Public Radio. Retrieved from https://www.npr.com

16
Happiness as a Life Goal

Almost everyone enjoys being happy. But is it something we should perpetually aspire to? The American Declaration of Independence entitles us to the "pursuit of happiness." Pursuit is a quest. Happiness, from the same root as "happen" and "perhaps," means luck ("Happiness," 2015). We have a right to hope for good circumstances – not exactly a high standard. Similarly, psychology focuses happiness on feeling good and evaluating our lives as "good enough" (Diener, 2000). Again, not a high standard.

A children's song repeats: "If you're happy and you know it, clap your hands!" Researchers have come up with more assessment possibilities: Is happiness a momentary feeling (Lyubomirsky, King, & Diener, 2005), a self-evaluation (Diener, Emmons, Larsen & Griffin, 1985), making meaning of events (Haybron, 2011; James, 2000), or reasoning well for a greater good (Shields, 2014)? Is it a benefit in a cost-benefit accounting of life (Helliwell, Layard, & Sachs, 2015), or an achievement of what we expect (Shute, 2014), or a comparison of our lives to others' (Hogenboom, 2014)? Is happiness built from big dreams or small moments (Brooks, 2015), great achievements or everyday tasks well done (Cohen, 2015)?

Each of these frames on happiness offers different ethical opportunities and questions about whether happiness should be placed on the pedestal of "life goal." Individually and collectively, is it good to pursue happiness? Or is it something we should frown upon?

The Smiley Face Reigns

with Robert Kagan

The word "happy" adjusts its meaning to societal, cultural, and historical milieu (McMahon, 2006; Weiner, 2008). In Western history, Greek mythology portrays it as the gods' gift for a chosen few. Until Socrates, no one thought to strive for happiness. Christians placed happiness mostly in the afterlife. Enlightenment thinkers made it an inalienable right. Modern psychologists tie happiness to pleasure, meaning, and purpose (James, 2000; Ryan & Deci, 2001). A "tour" of 10 contemporary countries found happiness associated with money and good luck, tolerance of diversity, well-run routines, even binge drinking (Weiner, 2008). Increasingly, happiness is being adopted as a national public policy measure, akin to the GDP, to compare well-being across countries over time (Helliwell et al., 2012, 2013, 2015; Novotney, 2014). Today, with a branch of Western social science dedicated to happiness, Positive Psychology, happiness is considered pleasant feeling, life satisfaction, and sense that "I'm doing ok" (Diener, 2000).

Does money buy happiness?

For most of history, most people toiled (McMahon, 2006). They sought survival, not happiness. Even today, to some degree, making a living is foundational to making a happy life. When asked about life goals, young Americans focus first on work and career (Moran, 2014). In whatever field they choose, they expect the work to have interest, quick promotions, high pay, and time for other life pursuits (Ng, Schweitzer, & Lyons, 2010).

Yet, numerous studies suggest that money goes only so far toward happiness (e.g., Diener & Oishi, 2000). In times of scarcity, acquiring another dollar improves happiness (Mullainathan, 2012). But after making enough to live moderately well, another dollar won't add that much. Still, buying "stuff" is what happiness is made of (BBC Staff, 2015; Quartz & Asp, 2015). Materialistic happiness may become problematic as we perpetually seek more, more, more (Diener, 2000). The more pressure to be happy, the more anxiety about whether we are happy enough, which may cause exhaustion or depression (BBC Staff, 2015). And materialistic happiness can waste resources, develop addictions, contribute to landfills, and breed inequality (e.g., Porter, 2015).

People who define happiness in economic terms play a zero-sum game, competing *against* others for a "pot of gold" that only some will reach (Cowen, 2015). This game may lead to a winner-take-all society that ends up as a no-win situation. The incessant pursuit of profit by the 1%ers, at the expense of the middle class, may leave wealthy executives and investors alone, as insufficient consumers remain who can buy their products and services (see Klein, 2015).

What do we target?

Happiness stems from comparisons – to our own expectations and to others' statuses. First, we imagine what we want and reap good feelings dreaming about how great it will be. If we pursue but don't acquire it, disappointment sets in. If we do acquire it, we feel excited. Then we adapt, and crave "what's next?" If we set goals high, we stretch more and increase potential to fail, but we enjoy the anticipation (Shute, 2014). If we set goals low, we thwart disappointment but may enjoy the victory less (Hogenboom, 2014). Second, we are happier when we are doing better than others (Hogenboom, 2014). Nothing makes this more apparent than social media. Concocting our own "curated selves" and consuming the curated selves of others can make us miserable (A. Brooks, 2014).

The more happiness we pursue, the more we demand, breeding restlessness for an ever-present emotional "high" that may not be personally, economically, or socially sustainable (McMahon, 2006). If real-life experience can't meet our happiness demands, or we don't want to expend the effort, then medicine might help. Pharmaceutical companies have profited from drugs that have become lifestyle enhancers or "cosmetics" for our emotions (McMahon, 2006).

Is happiness a new cultural imperialism?

Citizen happiness has been promoted as a new metric for governments, akin to gross domestic product (Novotney, 2014). Surveys poll citizens' emotional reactions to their country's employment, communal values and trust, government effectiveness, health, life balance, safety, and education (Helliwell et al., 2012).

Although Bhutan was the first country, in 1972, to declare a gross national happiness initiative (Helliwell et al., 2012; Weiner, 2008), and it has rallied the United Nations and other countries to follow suit (Ryback, 2012), the scientific foundations of these international developments have a Western flavor (Diener, 2000). Happiness, based on American individualism, may become the next horizon for cultural imperialism.

Collective goals, including for citizen happiness, create narratives of ideal ways to live. Governments may create activities, events, and programs to induce happiness. Or they may "mandate" happiness through economic incentives, tax penalties, perhaps behavior modification, or even pharmaceutical intervention. But populations are rarely homogeneous these days. Those with the most power may have more sway over these ideal ways. National happiness goals may make unhappy some citizens who have different goals. Unhappy people cause trouble – arguments, protests, petty crime, rioting. A government may elect to remove unhappy people by force to maintain a high "average" happiness.

Happiness breeds contentment, then complacency

Happiness is generally good for a collective. Happy adults live longer and contribute to community initiatives (Diener, 2000). Happy workers are more productive, creative, and cooperative (Helliwell et al., 2013). It seems like a win-win situation. But these bright-eyed morale boosters may have a dark side: the contentment they induce may morph into complacency. Entertaining events, giveaways, free treats, and other relatively inexpensive incentives focus attention on visible, short-term pleasures. They also may blind us to aspects of life that enhance chances for long-term happiness (Quartz & Asp, 2015). For example, our current overabundance of choices in nearly every area of life may occupy our time and our mind so we don't realize how wages stagnate, social connections loosen, and equality becomes precarious. These trends are more likely to affect our capability to be happy in the future because they are the forces that shape later opportunities.

Happiness can also lead to passivity. Especially if happiness becomes a national goal of governments, a "service" our leaders provide to us, we may relinquish our ability to make ourselves happy. These bureaucratic programs may be expensive, difficult to sustain, and perhaps ineffective if they follow conventional, short-term "morale boosting" initiatives. Plus, once government programs are in place, we learn to expect those benefits and become *entitled* to them. Then they lose the power to generate further happiness, but can make us unhappy if they are removed.

Happiness may lower our ability to cope and thrive

Happiness often is promoted as the "easy road." When we are happy, our needs are met. We don't seek anything or anyone. But happiness as contentment can start to feel like nothing. We don't like emptiness: one study showed people prefer shocks to facing the emptiness inside their own minds (Webb, 2014). Suffering has benefits.

We gather when we need each other: to solve problems, address challenges, negotiate issues, or mark turning points like weddings, funerals, and graduations. These encounters come with suffering through the issue. Uncertainty, trauma, difficulties pull us together into a community, show us we are part of something larger than ourselves, and provide opportunity to transcend the here-and-now. Perhaps what is important in the "pursuit of happiness" is the pursuit, not the happiness. The seeking, not the finding, provides benefit (A. Brooks, 2014; Helliwell et al., 2012). Making progress is the source of gratification – not immediately, but with time and effort (Diener, 2000).

One characterization of happiness in the ancient world was the "tragic hero" who beat the odds of external circumstances or internal conflicts (McMahon, 2006). Another more modern characterization is the "happy ending," a reflection on one's life that concedes not every moment was pleasant, but all's well that ends well (McMahon, 2006). In both of these characterizations, most significant and appreciated are the challenges that forged our character (D. Brooks, 2014). Dissatisfaction does not equal unhappiness, and such agitation could be helpful to long-term and/or collective happiness (Alessandri, 2014).

Selfish gratification or higher calling?

Researchers separate happiness into two types (Ryan & Deci, 2001). Hedonic happiness is satisfying personal interests and pleasures through material possessions, thrilling experiences, and avoidance of responsibility. It focuses on self-oriented gratification. Hedonic happiness focuses on what we *get* out of life.

Eudaimonic happiness is finding meaning or significance in life events and pursuing higher callings beyond personal gratification, for example, through serving others (Smith, 2013). It focuses on contribution to a common good. Eudaimonic happiness focuses on what we *give* to life. Eudaimonic happiness is not a zero-sum game. Everyone could benefit so well-being for all can rise (Helliwell et al., 2013).

The Millennial generation seems amenable to both hedonistic and eudaimonic happiness pursuits. On one hand, compared to prior generations, their job search seems more about personal satisfaction of their interests (Alsop, 2008). On the other hand, they desire to make an impact on the community, and more graduates are looking for careers in non-for-profit and social entrepreneurship (Rampell, 2011; Shapira, 2008).

Conventional wisdom separates particular jobs into more or less meaningful, which suggests that some positions critical to societal well-being but that don't inherently bring happiness or meaning, like sanitation, may be avoided. But Millennials' focus on personal meaning may allow a happy disposition regardless of position. For example, a dancing toll taker demonstrated how meaning is given to a situation, not extracted from it. Although his job in a highway tollbooth might be considered dull, he construed it as having a great view of the Golden Gate Bridge and providing a venue to make sure the drivers who came through received at least one smile a day (Garfield, 1987). He was happy, and in his humble way he made others happy.

Perhaps happiness depends on how we look *beyond* ourselves. If we expect the world to provide us incentives or rewards to make us happy, we may be setting ourselves up for unhappiness. But if we consider happiness a contribution to the world, we may make everyone happier. Studies show that even monkeys and children who are kind have more friends, fun, and happiness than self-pleasers (BBC Staff, 2015). Happiness may come down to valuing relationships with others (Brooks, 2013; Helliwell et al., 2015).

Perhaps the most poignant example of this self-other distinction is parenthood. Although many people report it is one of the most meaningful experiences in life, parenting often ranks low in hedonic happiness (Hansen, 2012; Lyubomirsky & Boehm, 2010). It's not fun sometimes, and parents can worry and suffer a lot. If adults only thought of their own self-gratification, there probably would be fewer babies born. Children provide long-term benefits at the expense of short-term pleasures. Looking back, most parents wouldn't have it any other way.

Further exploration

1. What is your personal definition of happiness? Within your list of life goals, where does happiness rank? What criteria does your definition provide for you to "measure" how happy you are – how well you are achieving the goal?
2. Design a society focused *entirely* on happiness. It has no other goals for its people. What institutions would be most important, and why? What type of government would be appropriate, and why? What activities would occupy people's time? How could you make this society sustainable?
3. To what extent should societies focus on the happiness of its citizens? Rank happiness as a priority compared to the economy, defense, education, the arts, technology, and other societal interests. Defend your ranking.

References

Alessandri, M. (2014, November 23). Companions in misery. *The New York Times*, p. SR4. Retrieved from https://www.nytimes.com
Alsop, R. (2008). *The trophy kids grow up: How the millennial generation is shaking up the workplace*. San Francisco, CA: Jossey-Bass.
BBC Staff. (2015, January 2). The pursuit of happiness. British Broadcasting Corporation. Retrieved from https://www.bbc.com
Brooks, A. C. (2013, December 14). A formula for happiness. *The New York Times*, p. SR1. Retrieved from https://www.nytimes.com
Brooks, A. C. (2014, July 20). Love people, not pleasure. *The New York Times*, p. SR1. Retrieved from https://www.nytimes.com
Brooks, D. (2014, April 7). What suffering does. *The New York Times*, p. A25. Retrieved from https://www.nytimes.com
Brooks, D. (2015, May 29). The small, happy life. *The New York Times*. Retrieved from https://www.nytimes.com
Cohen, R. (2015, June 12). Mow the lawn. *The New York Times*. Retrieved from https://www.nytimes.com
Cowen, T. (2015, April 5). It's not the inequality; it's the immobility. *The New York Times*, p. BU6. Retrieved from https://www.nytimes.com
Diener, E. (2000, January). Subjective well-being: The science of happiness and a proposal for a national index. *American Psychologist, 55*(1), 34–43.
Diener, E., Emmons, R. A., Larsen, R. J., & Griffin, S. (1985). The Satisfaction with Life Scale. *Journal of Personality Assessment, 49*(1), 71–75.
Diener, E., & Oishi, S. (2000). Money and happiness: Income and subjective well-being across nations. In E. Diener & E. M. Suh (Eds.), *Culture and subjective well-being* (pp. 185–218). Cambridge, MA: MIT Press.
Garfield, C. (1987). *Peak performers: The new heroes of American business*. New York, NY: William Morrow.
Hansen, T. (2012). Parenthood and happiness: A review of folk theories versus empirical evidence. *Social Indicators Research, 108*(1), 29–64.
Happiness [Def. 1a, 1b]. (2015). In *Oxford English Dictionary Online*. Retrieved August 21, 2015, from http://www.oed.com
Haybron, D. (2011, Fall). Happiness. In E. N. Zalta (Ed.), *The Stanford encyclopedia of philosophy*. Retrieved from http://plato.stanford.edu
Helliwell, J. Layard, R. & Sachs, J (2012). *World happiness report*. New York, NY: Earth Institute.
Helliwell, J. Layard, R. & Sachs, J (2013). *World happiness report 2013*. New York, NY: Earth Institute.
Helliwell, J. Layard, R. & Sachs, J (2015). *World happiness report 2015*. New York, NY: Earth Institute.
Hogenboom, M. (2014, August 4). Equation "can predict momentary happiness." British Broadcasting Corporation. Retrieved from https://www.bbc.com
James, W. (2000). *Pragmatism and other writings* (Giles Gunn, ed.). London: Penguin Books.
Klein, N. (2015, March 22). Greed is good, for some [Book review of "The age of acquiescence," by Steve Fraser]. *The New York Times*, p. BR12. Retrieved from https://www.nytimes.com

Lyubomirsky, S., & Boehm, J. K. (2010). Human motives, happiness, and the puzzle of parenthood: Commentary on Kenrick et al. (2010). *Perspectives on Psychological Science, 5*(3), 327–334.
Lyubomirsky, S., King, L., & Diener, E. (2005). The benefits of frequent positive affect: Does happiness lead to success? *Psychological Bulletin, 131*(6), 803–855.
McMahon, D. M. (2006). *Happiness: A history.* New York: NY: Grove Press.
Moran, S. (2014). What "purpose" means to youth: Are there cultures of purpose? *Applied Developmental Science, 18*(3), 1–13.
Mullainathan, S. (2012). Psychology and development economics. In P. Diamond & H. Vartiainen (Eds.), *Behavioral economics and its applications* (pp. 85–113). Princeton, NJ: Princeton University Press.
Ng, E. S. W., Schweitzer, L. & Lyons, S. T. (2010). New generation, great expectations: A field study of the millennial generation. *Journal of Business and Psychology, 25*(2), 281–292.
Novotney, A. (2014, March). Gross national well-being. *Monitor on Psychology,* pp. 23–24.
Porter, E. (2015, April 29). Income inequality is costing the U.S. on social issues. *The New York Times,* p. B1. Retrieved from http://www.nytimes.com
Quartz, S., & Asp, A. (2015, April 12). Unequal, yet happy. *The New York Times,* p. SR4. Retrieved from http://www.nytimes.com
Rampell, C. (2011, May 1). More college graduates take public service jobs. *The New York Times.* Retrieved from http://www.nytimes.com
Ryan, R. M., & Deci, E. L. (2001). On happiness and human potentials: A review of research on hedonic and eudaimonic well-being. *Annual Review of Psychology, 52,* 141–166.
Ryback, T. W. (2012, March 28). The U.N. Happiness Project. *The New York Times.* Retrieved from https://www.nytimes.com
Shapira, I. (2008, October 14). For this generation, vocations of service. *The Washington Post.* Retrieved from http://www.washingtonpost.com
Shields, C. (2014, Spring). Aristotle. In E. N. Zalta (Ed.), *The Stanford encyclopedia of philosophy.* Retrieved from http://plato.stanford.edu
Shute, N. (2014, August 6). Do you want to be happy? Don't set your expectations too high [Shots web log post]. National Public Radio. Retrieved from https://www.npr.org
Smith, E. (2013, January 9). There's more to life than being happy. *The Atlantic.* Retrieved from http://www.theatlantic.com
Webb, J. (2014, July 4). Do people choose pain over boredom? British Broadcasting Corporation. Retrieved from https://www.bbc.com
Weiner, E. (2008). *The geography of bliss.* New York, NY: Twelve.

17
Boredom Avoidance

"I'm bored!" It's that dreaded feeling of...nothingness. Our basic physical needs have been met so we don't need anything from the environment, but...something please happen. Long stretches of boredom are endemic in some important jobs often associated with excitement – like policing, firefighting, and space travel (Aschwanden, 2015). A bored person is disconnected from the situation (Weir, 2013), is not valuing the surroundings (Iso-Ahola & Weissinger, 1987), and can't figure out a way to perceive the setting as opportunities to participate (Hamilton, Haier, & Buchsbaum, 1984). And now boredom has become a psychological topic worthy to measure (Hunter, Dyer, Cribbie, & Eastwood, 2015).

Marketers' incessant persuasion that better lives are found in their offerings – entertainments, gadgets, "experiences" – creates an amplifying feedback loop of needing more and more external stimulation. Once we've habituated and we are comfortable about what comes next, life loses its suspense, its surprise (Ely, Frankel, & Kamenica, 2015). If life is not exciting, we might be missing out. Or worse: to be bored is to be boring, we fear. "I can't stand it!" For some people, the discomfort from spending time devoid of external stimulation is shocking (Wilson, Reinhard, Westgate, Gilbert, et al., 2014).

What does this constant search for stimulation do to our bodies, minds, relationships, and social institutions? Are we medicalizing this disconnect: is attention deficit hyperactivity disorder an extreme form of boredom (Friedman, 2014)? Why are we spending so much effort – and money (entertainment is big business) – to avoid an emptiness rather than generating momentum toward what could fill that emptiness, such as daydreaming, curiosity, interest, and perhaps creativity? Is boredom avoidance a sustainable value?

Like Watching Paint Dry
with Kaitlin Black

What is going on when nothing is going on? The experience of boredom probably goes back to when humans figured out how to take care of their basic needs and developed "free time." But, the word itself is more contemporary. According to the Oxford English Dictionary, the word "boredom" first appears in 1853 in Charles Dickens' *Bleak House* ("Boredom," 2013).

Among the first to opine on boredom and why it should be avoided were the ancient Romans. The philosopher Seneca believed that boredom in its extreme could take over a person's life, to the point where suicide is the only logical conclusion (Toohey, 2004). The historian Plutarch's account of the life of the monarch Pyrrhus describes the leader as being bored "to the point of nausea" in his retirement (Toohey, 1987, p. 199).

Centuries later, the Christians categorized boredom, or *acedia*, as one of the eight cardinal sins (Jackson, 1985). Literature during the Renaissance and the Enlightenment captured both the feelings of ennui and nausea (Leroux, 2008). Thus, throughout history, boredom has been viewed as something to avoid. Yet, boredom also could be considered a luxury since there was no time to be bored if life was spent struggling to survive and make a living (Weir, 2013).

A growing focus on boredom

Beginning in the 1930s, boredom became a topic of scientific investigation. Psychologist Joseph Ephraim Barmack (1939) found that stimulants reduced reports of boredom and increased levels of attention in college students. Since the 1960s, organizational psychologists focused on boredom's role in undesirable work behaviors (Bruursema, Kessler, & Spector, 2011; cf., Jackson, Masso, & Vadi, 2013), decreased job performance, and increased dissatisfaction and absenteeism (Kass, Vodanovich, & Callender, 2001).

Today, boredom is viewed as "the aversive experience of wanting, but being unable, to engage in satisfying activity" (Eastwood, Frischen, Fenske, & Smilek, 2012, p. 483), as a lack of movement or purpose (Brissett & Snow, 1993). Boredom correlates with misbehaving (Weir, 2013) and crime (Goodman, 2014), as well as a range of mental health disorders, such as depression (Schaeffer, 1988), anxiety (Fahlman, Mercer, Gaskovski, Eastwood, & Eastwood, 2009), addiction (Nichols & Nicki,

2004), and overeating (Abramson & Stinson, 1977). Because boredom is associated with so many negative outcomes, perhaps it should be avoided.

More recently, boredom as a launching pad to creativity and positive change has been explored. People generally want to remain in a state of optimal arousal (Nakamura & Csikszentmihalyi, 2009): If a person's skills exceed the task, they will quickly become bored. They will adjust their skill level or the challenge level to return to optimal arousal, or "flow." We seem to be "wired" to avoid boredom.

The ethical dimensions of avoiding boredom can be envisioned as a ripple of effects, starting with an individual then moving into social fields and the broader culture.

What do boredom and its alleviation feel like?

Boredom is felt as lethargy and alienation from one's own emotions (Weir, 2013). An empty mind is breeding grounds for negative thoughts to creep in, so we prefer to distract ourselves with external stimulation rather than sit with our own dull inner lives (Murphy, 2014; Webb, 2014). What's more, not being seen as an "engaging personality" – that is, being considered a "boring person" – has nearly become a character flaw not just a temporary emotional state. Yet, boredom creates its own agitation and frustration to seek relief. Most people will do almost anything to avoid this feeling – or *lack* of feeling – including shocking oneself just to have some stimulation (Wilson et al., 2014).

Avoiding boredom comes at a cost. Hyper-scheduled, digitally mediated lifestyles can lead to information overload and distraction addiction (Dokoupil, 2012; Morozov, 2013). Although gadgets might seem to alleviate boredom, these devices may increase the prevalence of boredom because people do not learn to occupy themselves (Weir, 2013). It is another way that smartphones may be replacing, rather than augmenting, human minds.

There could be physical consequences to constant stimulation and failure to disconnect for a while. Increased heart rate, elevated blood pressure, and neurological disorders could become more common, even among age groups where it is typically nonexistent, like young adults and children. We may see future generations dying younger as we increasingly turn to the next thrill.

Due to ever-increasing stimulation, sleep disorders could arise. Even in the few moments before nodding off, many people check their electronic devices, which may interrupt the ability to fall asleep (Bilton,

2014). Exhaustion comes with consequences. More people drive drowsy and are less productive at work. Plus, they may not have the energy to enjoy the activities they do engage. Workers in dangerous jobs could harm themselves or others if they are not alert.

Psychological disorders may increase. As emotional attachment to electronic devices grows (Yu, 2014), people may suffer separation anxiety or even mental breakdowns when the devices fail. These techno-buddies become a primary source of all emotional stimulation as our social interactions become increasingly mediated. Eventually, without them, perhaps we feel nothing – at the extreme, no emoticons, no emotions. We become disconnected from our bodies, our visceral *feeling* of emotions. Although this digitally mediated world is the norm, there are pockets of opportunity to unplug. People may try to cultivate mindfulness or stillness as a virtue.

Of course, a pharmaceutical solution to boredom may develop (some might say legalized marijuana is a current solution). If boredom becomes something to cure not just avoid, watch for medications touted as boredom-busters. Ideally, if boredom is eradicated, associated ailments could be improved as well. Expensive therapies used to treat depression and anxiety give way to a pill that removes the root cause of boredom. However, an anti-boredom medication might increase the risk of these other ailments, particularly if it is similar to other stimulants currently available, such as caffeine or hallucinogens.

What if some people like to "space out," but an efficiency-driven society deems that inappropriate behavior – perhaps criminal (like "embezzling" time in unproductive activities)? Doctors, employers, and judges may mandate that these misfits take a neuro-enhancing "pep pill" for the good of others, the company, or society (Hall, 2004). Then, these people could lose control over their bodies and their subjective experiences. In the future, we might see court cases determining whether whole countries should strive to avoid boredom.

How does avoiding boredom affect social connections?

Boredom is not just an individual emotional issue. It creates social challenges as well (Iso-Ahola & Weissinger, 1987). Because boredom defines a particular kind of relationship with the environment (Weir, 2013), trying to avoid boredom can alter relationships with other people and tasks (Friedman, 2014). Absent the boring downtime to reflect, people tend to be less empathetic. Becoming out of touch with one's own emotions inhibits connecting with others (Murphy, 2014). Genuine conversations may require some "boring bits" as participants muddle through topics, comments, pauses, and interruptions to discover the "interesting bits" (Garber, 2014). Tolerance for real conversations may fall because they aren't efficient, scripted, edited like movies or texting (Garber, 2014; Turkle, 2011).

It is in the best interest of advertisers and the media to keep people from getting bored. As more people watch shows on-demand, alone whenever they're bored, instead of getting together to watch a scheduled program, friends who bond over a favorite show may not be able to talk about the most recent episode without creating spoilers. Sharing is reduced.

How might social institutions adapt if society goes "boredom-free"?

Institutions are stable ways for people to interact with each other. They are designed to be boring in their constancy so we can rely on them. Generally, we want our schools, companies, and government to be efficient. But as we strive to avoid boredom, that reliability could become problematic.

Classrooms may shift to grouping students by complementary ability and interests, not by age, to keep them optimally stimulated (Larson & Richards, 1991). Personalized curricula could replace current curricula that target the fictional "average" student, which results in most students being a little bored. Educators may invent class styles that require students to contribute rather than listening passively. Students engaging environmental challenges (Nakamura & Csikszentmihalyi, 2009) – the opposite of boredom (Weir, 2013) – becomes fundamental to academic success. Rather than standardizing lessons or experiences, designing in unpredictability (Friedman, 2014) to generate suspense and surprise (Ely et al., 2015) in learning becomes the norm. Tailored education comes not from providing more services *to* students, but expecting more *from* them.

Workplaces may need to be "gamified" so that employees are motivated to seek adventures and rewards in their jobs (Wingfield, 2012). Of course, this may be a losing battle. As the level of stimulation necessary to avoid boredom rises, employers and jobs may have to become even more entertaining as workers come to expect the external stimulation. This might mean booming business for entertainment companies – not only to entertain consumers but as consultants to turn jobs into "thrill rides."

During leisure time, more people could look to cutting-edge, death-defying experiences: old thrills like jumping out of airplanes or climbing mountains may give way to space tourism, even if it comes with no return trip (Klosterman, 2015). Most likely, technology companies will continue to profit from boredom avoidance in an accelerating race of excitement, even if it is virtual and not real.

It is in a government's best interest to keep citizens engaged. Bored people have time to think about their circumstances, to compare themselves to others, and to question policies. This could cause dissatisfaction, and if it festers, lead to protests and calls for change. Officials could risk lost reelections. Subsidizing activities to placate the bored are likely to continue, just as cities today create events on weekends and holidays so people can have something to focus on. Officials could create contests so that citizens could create opportunities to alleviate each other's boredom in legal ways rather than considering illegal activities as exciting.

Can boredom be necessary or even desirable?

At a time when overstimulation and fast-paced lifestyles are the norm, at some point the tide may change and some people turn to boredom to slow themselves down. People are starting to converge on different forms of "unplug-a-thons." A public radio station sponsored a one-week Bored and Brilliant Challenge to encourage listeners to seek "mental down time" (Cornish, 2015). With the emergence of popular conferences such as *Boring*, where people willingly listen to dull talks that turn tedious topics into something interesting (Naik, 2010), and Internet video sites dedicated to sharing videos of mundane life (Wortham, 2015), boredom might become something to embrace. People get in touch with themselves and their immediate environment, not just stay in constant touch through devices.

In some ways, boredom creates its own demise. If allowed to last for a period of time, boredom's emptiness makes room for interests. The ethical implications above assume that stimulation must come from outside the person. Studies of creativity suggest that boredom may be a precursor to breakthrough thinking: boredom generates motivation to seek intellectual engagement, meaning, and purpose from *within* one's own mind (Weir, 2013). We become self-generators of what excites us.

Boredom opens the mind to new possibilities that would be crowded out by external distractions (Baird, Smallwood, Mrazek, Kam, et al., 2012; Morozov, 2013). Perhaps that's why so many historical creators mention being sickly children. Having to stay in bed for long periods forced them to develop rich inner lives and imaginations. Famous creators often talk of boredom as a step in the creative process – there is a *need* to waste time and loaf to increase mental flexibility (Mann, 2015). Some of the greatest innovations were born through zealous repetition to discover something unusual and valuable (Davidson, 2014). Perhaps creativity's "aha!" moment is the payoff for the tedium of bringing a truly novel idea into reality. If we avoid boredom, then, are we inadvertently thwarting the introduction of new ideas and, thereby, slowing cultural development?

Further exploration

1. You work for an insurance company and are charged with determining which potential clients are risky for your company to take on. Design a questionnaire that determines their boredom levels, both present and projected into the future.
2. Imagine the United Nations has decreed that boredom should be eradicated as part of a 50-year global plan to increase happiness. Draft a report to your government outlining specific strategies plus potential problems that those strategies could cause.
3. Your elderly mother shows symptoms of boredom. A new drug that reduces boredom has gained FDA approval, and your mother's doctor wants to put her on the medication. What is your opinion? What do you consider the most important issues to consider?

References

Abramson, E. E., & Stinson, S. G. (1977). Boredom and eating in obese and non-obese individuals. *Addictive Behaviors, 2,* 181–185. doi: 10.1016/0306-4603(77)90015-6
Aschwanden, C. (2015, May 11). Review: In "Extreme," by Emma Barrett and Paul Martin, psychologists explore those who test limits [Book review]. *The New York Times.* Retrieved from http://www.nytimes.com
Baird, B., Smallwood, J., Mrazek, M. D., Kam, J. W. Y., Franklin, M. S., & Schooler, J. W. (2012). Inspired by distraction: Mind wandering facilitates creative incubation. *Psychological Science, 23,* 1117–1122. doi: 10.1177/0956797612446024
Barmack, J. E. (1939). Studies on the psychophysiology of boredom: Part I. The effect of 15 mgs. of benzedrine sulfate and 60 mgs. of ephedrine hydrochloride on blood pressure, report of boredom, and other factors. *Journal of Experimental Psychology, 25,* 494–504. doi: 10.1037/h0054402
Bilton, N. (2014, February 9). Disruptions: For a restful night, make your smartphone sleep on the couch [Bits web log]. *The New York Times.* Retrieved from http://bits.blogs.nytimes.com

Boredom [Def. 2]. (2013). In *Oxford English Dictionary Online*, Retrieved October 15, 2013, from http://www.oed.com

Brissett, D. & Snow, R. P. (1993). Boredom: Where the future isn't. *Symbolic Interaction, 16*, 237–256. doi: 10.1525/si.1993.16.3.237

Bruursema, K., Kessler, S. R., & Spector, P. E. (2011). Bored employees misbehaving: The relationship between boredom and counterproductive behaviour. *Work & Stress, 25*, 93–107. doi: 10.1080/02678373.2011.596670

Cornish, A. (2015, February 2). It's time to get bored – and brilliant [All Tech Considered web log post]. National Public Radio. Retrieved from https://www.npr.org

Davidson, A. (2014, November 16). Losers win. *The New York Times Magazine*, pp. 40–50.

Dokoupil, T. (2012, July 18). Is the onslaught making us crazy? *Newsweek*, pp. 24–30.

Eastwood, J. D., Frischen, A., Fenske, M. J., & Smilek, D. (2012). The unengaged mind: Defining boredom in terms of attention. *Perspectives on Psychological Science, 7*(5), 482–495.

Ely, J., Frankel, A., & Kamenica, E. (2015, April 26). The mathematics of suspense. *The New York Times*, p. SR9. Retrieved from http://www.nytimes.com

Fahlman, S. A., Mercer, K. B., Gaskovski, P., Eastwood, A. E., & Eastwood, J. D. (2009). Does a lack of life meaning cause boredom? Results from psychometric, longitudinal, and experimental analyses. *Journal of Social and Clinical Psychology, 28*(3), 307–340. doi: 10.1521/jscp.2009.28.3.307

Friedman, R. A. (2014, October 31). A natural fix for A.D.H.D. *The New York Times*. Retrieved from https://www.nytimes.com

Garber, M. (2014, January/February). Saving the lost art of conversation. *The Atlantic*. Retrieved from http://theatlantic.com

Goodman, J. D. (2014, April 8). Teenager tells police boredom led him to start fire that injured 2 officers. *The New York Times*, p. A22. Retrieved from https://www.nytimes.com

Hall, W. (2004, December). Feeling "better than well." *EMBO Reports, 5*, 1105–1109. doi: 10.1038/sj.embor.7400303

Hamilton, J. A., Haier, J. R., & Buchsbaum, M. S. (1984). Intrinsic enjoyment and boredom coping scales: Validation with personality, evoked potential and attention measures. *Personality and Individual Differences, 5*, 183–193. doi:10.1016/0191-8869(84)90050-3

Hunter, J. A., Dyer, K. J., Cribbie, R. A., & Eastwood, J. D. (2015). Exploring the utility of the Multidimensional State Boredom Scale. *European Journal of Psychological Assessment*. Advance online publication. doi:10.1027/1015-5759/a000251

Iso-Ahola, S. E., Weissinger, E. (1987). Leisure and boredom. *Journal of Social & Clinical Psychology, 5*, 356–364. doi: 10.1521/jscp.1987.5.3.356

Jackson, K., Masso, J., & Vadi, M. (2013). The drivers and moderators for dishonest behavior in the service sector. In M. Vadi & T. Vissak (Eds.), *(Dis)honesty in management: Manifestations and consequences* (pp. 169–193). London, UK: Emerald.

Jackson, S. W. (1985). Acedia the sin and its relationship to sorrow and melancholia. In A. Kleinman & B. Good (Eds.), *Culture and depression: Studies in the anthropology and cross-cultural psychiatry of affect and culture* (pp. 43–62). Los Angeles, CA: University of California Press.

Kass, S. J., Vodanovich, S. J., & Callender, A. (2001). State-trait boredom: Relationship to absenteeism, tenure, and job satisfaction. *Journal of Business and Psychology, 16*, 317–327. doi: 10.1023/A:1011121503118

Klosterman, C. (2015, February 8). The hazards of other planets. *The New York Times Magazine*, p. MM17. Retrieved from https://www.nytimes.com

Larson, R. W., & Richards, M. H. (1991). Boredom in the middle school years: Blaming schools versus blaming students. *American Journal of Education, 99*(4), 418–443.

Leroux, J. (2008). Exhausting ennui: Bellow, Dostoevsky, and the literature of boredom. *College Literature, 35*(1), 1–15.

Mann, M. (2015, April 19). The other side of boredom. *The New York Times*, p. SR6. Retrieved from http://www.nytimes.com

Morozov, E. (2013, October 28). Only disconnect: Three cheers for boredom. *The New Yorker*, pp. 33–37.

Murphy, K. (2014, July 25). No time to think. *The New York Times*. Retrieved from https://www.nytimes.com

Naik, G. (2010). Boredom enthusiasts discover the pleasures of understimulation. *The Wall Street Journal*. Retrieved from http://online.wsj.com

Nakamura, J., & Csikszentmihalyi, M. (2009). Flow theory and research. In C. R. Snyder & S. J. Lopez (Eds.), *Handbook of positive psychology, 2nd ed.* (pp. 195–206). New York, NY: Oxford University Press.

Nichols, L. A., & Nicki, R. (2004). Development of a psychometrically sound internet addiction scale: A preliminary step. *Psychology of Addictive Behaviors, 18*(4), 381–384.

Schaeffer, N. C. (1988). An application of item response theory to the measurement of depression. *Sociological Methodology, 18*, 271–307.

Toohey, P. (1987). Plutarch, Pyrrhus. 13: άλνςναντιώδης, *Glotta* 65 (pp. 199–202). Göttingen, Germany: Vandenhoeck & Ruprecht.

Toohey, P. (2004). *Melancholy, love, and time: Boundaries of the self in ancient literature*. Ann Arbor, MI: University of Michigan Press.

Turkle, S. (2011). *Alone together: Why we expect more from technology and less from each other*. New York, NY: Basic Books.

Webb, J. (2014, July 4). Do people choose pain over boredom? British Broadcasting Corporation. Retrieved from https://www.bbc.com

Weir, K. (2013, July/August). Never a dull moment. *Monitor on Psychology*, pp. 54–57.

Wilson, T. D., Reinhard, D. A., Westgate, E. C., Gilbert, D. T., Ellerbeck, N., Hahn, C., Brown, C. L., Shaked, A. (2014, July 4). Just think: The challenges of the disengaged mind. *Science, 345*(6192), 75–77. doi: 10.1126/science.1250830

Wingfield, N. (2012, December 24). All the world's a game, and business is a player. *The New York Times*, p. A1. Retrieved from https://www.nytimes.com

Wortham, J. (2015, April 5). Borecore. *The New York Times*, p. MM16. Retrieved from http://www.nytimes.com

Yu, C. (2014, February 23). Happiness is a warm iPhone. *The New York Times*, p. SR1. Retrieved from http://www.nytimes.com

Part V
Self Definers

18
Authenticity as a Life Purpose

Who am "I"? Honest answers to that question evoke the authentic self. But which dimensions of me are true? My genes (Harmon, 2006)? My selfish desires (Critchley & Webster, 2013)? What I keep private inside my mind (Fehling, 2013)? My feelings (Harter, 2002)? The correspondence of my feelings and actions (Wood, Maltby, Baliousis, Linley, & Joseph, 2008)? Who I present to others (Seidman, 2013)? The "I" who continues from past to future (Hogenboom, 2014)? What "quantified self" gadgets measure (Singer, 2015)? Do I cultivate the true me on my own, or does the input of others help determine who I am (Kaplan, 2015; Luhrmann, 2014)? Will the "true me" please step forward?

Is "self-authenticity" a selfish "feel good" state or a more integrated trajectory that encompasses complexity, interconnection, and life meaning (Bauer, Schwab, & McAdams, 2011)? What happens if self-authenticity becomes everyone's life purpose, the ultimate goal for why each of us is here (Moran, 2014)? If everyone is focused only on themselves, is the "common good" a viable concept (Blow, 2014)? If community is still a worthy endeavor, how do we coordinate everyone's true selves?

True to My Self
with Olivia Lourie

The word of the year for 2013 was "selfie" (Turchi & Scalese, 2014). Self-help books and personal fulfillment programs have been around for decades. What is new is that "I" am always at the center of the picture (BBC Staff, 2015). Another recent development is the growing requirement for institutions to accommodate individual differences to an extreme (Blow, 2014). Some worry that people are losing the ability to bear situations that disappoint "me" (Critchley & Webster, 2013) or to recognize and contribute to a common good (Brooks, 2013).

From self-as-means to self-as-end

"Self" distinguishes "I" from others, allows continuity of today's and tomorrow's "I's," and harbors an inner world of thoughts and dreams (Markus & Kitayama, 1991). Self is the center of subjective experience. "Authenticity" means my inner feelings and outer expressions align and "I feel right" (Harter, 2002). "Life purpose" is a compass of something "I" consider important to steer my perceptions and behavior (Moran, 2014). The most common purposes focus on family and career (Moran, 2009). But Maslow's (1943) Hierarchy of Needs made the self, itself, into the compass by which we should direct our lives. At the top of his hierarchy is "self-actualization" – to reach our full potential. Our role in the world becomes self-referential – self-authenticity becomes life purpose.

Self and purpose as choices not dictates

For much of human existence, our lives were dictated by survival instinct, divine force, or social position. Then, 20th-century existential thinkers suggested it is *our* responsibility to define our self and our purpose (Crowell, 2015). Self and purpose became open questions full of uncertainty (Hookway & James, 2015; Olivares, 2010), and authenticity became the criterion to judge how well we fulfilled the responsibility to create a story for our self.

We become who we are with the help of guides and evaluators (Kaplan, 2015; Perry, 2014). Most cultures present a default self – a "starter kit" of perspectives and values appropriated from family, religion, education, government, and other institutions. Social sciences broadly categorize cultures into two camps with different default self configurations. Collectivist cultures best support an interdependent self who encompasses relations with others and aims to feel connected. Individualistic cultures best support an independent self who differentiates from others and aims to feel unique (Markus & Kitayama, 1991).

Recently, the Internet has presented tremendous variety in personae, lifestyles, and other options, so many people are no longer limited to their culture's defaults. These various ways of experiencing the world create both opportunities and challenges for self-authenticity and life purpose. What might be the ethical ripples of intentionally steering our lives by the compass to "be true to myself"? Since ethics addresses principles governing the effects of our actions on others, let's consider three ethical stances based on how different configurations of self relate to the common good.

Antagonistic: self versus the common good

John believes that those who differ from him are a threat, so he dedicates his life to fight their "wrongheaded" beliefs. He starts a website to educate people. It attracts many like-minded people. Their like-mindedness reinforces their current beliefs, which tends to keep the beliefs stable since no diversity is allowed. Individuals who start to think differently no longer visit the website.

In this ethical stance, the "I" and the "common good" (conceived as all others combined) are at war. The assumption is that I cannot be my authentic self without overcoming cultural pressures. The result is extreme individualism. If our life purpose is self-authenticity, then this stance requires us to reject what is outside our self. Those with different perspectives become trespassers on our life path.

These individuals could become detached from generally accepted purposes of life (Brooks, 2015). Shallower or no relationships may be sought. This independence would make consideration of others' rights or needs less prevalent: to each their own. If only one's own perspective counts, then negotiations, conversations, and collaborations would fail. Empathy, sympathy, trust, respect, partnership, friendship, love, and cooperation might become quaint, historical concepts. Eventually, there would be no "external pressures" to avoid because no one would interact. If no one interacts, there is no need for government – and democracy wouldn't function anyway. There is no common law to enforce – no rules, contracts, rewards, and punishments to which everyone agrees – because no one agrees.

Will this extreme individualism tear communities apart (Brooks, 2013)? The next generation, studies say, may not perpetuate the social institutions that have been considered the bedrock of society, such as marriage, family, religion, and politics (Pew Research Center, 2014). We are detached, going our own way without traditional, cross-generational guidance. I and my life are whatever I say they are, and no one else is allowed input: "You do you" and "I do me" with no overlap (Whitehead, 2015).

This stance could both increase and decrease extremism. On one hand, it can breed anomie, or social instability born of personal alienation from society. The authentic self is composed of restless alienation: to be "against" something outside and not "for" anything inside. Then, these individuals may end up being loners. However, alienation itself can be a magnet that gathers alienated individuals together to "circle the wagons" against the perceived external foe. If these individuals organize, then protest or violence may result.

Integrated: self as "the right person for the times"

Jim does what his family and friends do: sports and cooking. His life is comfortable. He abides by the laws, minds his manners, studies a popular major, and probably will join a prestigious company. He's grateful for the guidance of his coaches, teachers, and parents – it would be a lot of work to figure everything out himself. He's proud of his hometown. His quiet civilities may not be noted because that's how everyone is.

The integrated stance is extreme undifferentiation, with little distinction between the "I" and the "we." The assumption is that the undivided whole is the community, not the individual. The self is a collective effort, and one's authentic self is what society needs. The inner world mirrors the social world. Everyone's life purpose is a community-level purpose: to honor community, heritage, and family.

This ethical stance is more likely in collectivist cultures, such as in Asia (Markus & Kitayama, 1991). Although it may seem foreign in the West, it was not that long ago that the family was an indivisible unit. Work, marriage, and children were family, not individual, decisions to help the family survive and prosper (Hareven, 1991). This stance still prospers in some communities (Kuhn, 1948).

If self-authenticity is taken strictly in Maslow's individualistic terms, then this ethical stance may be at risk. Differences between individual and community goals may drive youth to reject or leave communities, who may not pass on the communitarian self to later generations. There is some evidence: A study of life purposes among Korean adolescents suggests that Western individualistic influences have increased personal gratification alongside the traditional focus on social responsibility (Shin, Hwang, Cho, & McCarthy-Donovan, 2014). However, if the self is the community, then this ethical stance may help alleviate lower civic engagement among youth (Malin, 2011). If the next generation increases identification with the communities that support them, and in turn, recognizes that preserving communities is a form of self-preservation, then communities might be rejuvenated. Personal gratification or disappointment is more quickly experienced, whereas community effects take time to materialize (Putnam, 2007), so communities that make salient the importance of cooperation may survive better.

Complementary: self as contributor to the common good

Sam has a passion for painting. Although he makes little money as an artist, he is satisfied (Hernholm, 2014). He likes how different viewers approach and interact with his pieces. His obvious joy encourages others to pursue their callings – whatever they enjoy, learn to master, and willingly share with others. He finds that their comments influence other viewers – like a domino effect of meaning-making – and they also sometimes inspire his own future pieces.

If the antagonistic stance is extreme individualism, and the integrated stance is extreme undifferentiation, then the complementary stance is a middle road – not in the sense of "average" but in the sense of "interaction." Neither the self nor the common good are predetermined. Rather, they fluidly compose each other through interaction. Communities are tapestries of differentiated selves (Putnam, 2007). One authentic self of subjectively defined interests, talents, and possibilities interweaves with other authentic selves to create the common good's "bigger picture" (Hernholm, 2014). An effective life purpose provides the self with an entry and pathway to contribute to the common good. As societies expand the possible roles that individuals may take, the common good becomes more complex (Kaplan, 2015).

This ethical stance encourages selves to pursue their talents as important for the common good (Moran, 2014). Life purpose's compass points toward neither personal gratification nor undifferentiated community participation, but toward personally significant contribution. A contribution mentality is not "either/or" – either I gain or you gain – but "yes, and" – if we both do our part, we both gain.

Predefining some roles as more valuable than others, as is often done in economics to rank jobs or fields in terms of salary or prestige, gives way to a recognition that diverse contributions are complementary to each other, and all are needed to make the common good viable and sustainable (Marino, 2014). For example, one often told story contrasts a bricklayer who defines himself as "a bricklayer," versus another bricklayer who views himself as "on the crew building a cathedral." Or another story describes a baseball batter whose talent is to load up the bases so his teammate can hit a homerun and bring in four runs: the homerun hitter's contribution may be more easily recognized, but the first batter's contribution is required to make the big plays. When individuals understand the larger role they can play in the common good, the effects can be amazing: even a highway toll taker can create a positive effect in the few moments he takes payment from drivers on their way to work (Garfield, 1987). A person's contribution, no matter how seemingly trivial, is valid. Fulfillment puts joy into the common good.

This ethical stance helps promote the contributions of creators and minority groups by changing our shared mindset from "mainstream versus different" to "symphony of voices." There is no "center" that is more worthy, but rather a fluidity of interacting contributions. Self-authenticity in this ethical stance asks us to step up to challenges and opportunities for which we can make a contribution, even if there is some risk, because we see how we matter to the overall common good. Mutual respect and understanding grows.

A concern of this ethical stance, however, is: what happens if a person's self is filled with negative energy and aims toward harmful contributions to society? This situation could happen if someone misconstrues the common good as a venue where "anything goes" or as a "market" for anything others will "buy." These construals misunderstand the common good, which is a shared resource to benefit the *whole* community, not individuals within the community (Sullivan, 2011). Thus, this ethical stance may not abolish the possibility of derogation or violence, and care must be taken to sensitize individuals to the ripple effects of their thoughts and actions.

Further exploration

1. What are possible distinctions between self-actualizing and being selfish? When might they overlap?
2. Imagine that you are part of a society in which everyone is purely self-motivating rather than responding to or contributing to shared societal or cultural ideas. How might this group of people maintain a cohesive culture?

References

Bauer, J. J., Schwab, J. R., & McAdams, D. P. (2011). Self-actualizing: Where ego development finally feels good? *The Humanistic Psychologist, 39*, 121–136. doi: 10.1080/08873267.2011.564978

BBC Staff. (2015, April 10). The tyranny of the selfie [A Point of View web log post]. British Broadcasting Corporation. Retrieved from https://www.bbc.com

Blow, C. M. (2014, March 7). The self(ie) generation. *The New York Times*, p. A19. Retrieved from https://www.nytimes.com

Brooks, D. (2013, June 11). The solitary leaker. *The New York Times*, p. A23. Retrieved from http://www.nytimes.com

Brooks, D. (2015, April 17). When cultures shift. *The New York Times*, p. A31. Retrieved from http://www.nytimes.com

Critchley, S., & Webster, J. (2013, June 30). The gospel according to "me" [Web log post]. *The New York Times*. Retrieved from http://opinionator.blogs.nytimes.com

Crowell, S. (2015, Spring). Existentialism. In E. N. Zalta (Ed.), *The Stanford encyclopedia of philosophy*. Retrieved from http://plato.stanford.edu

Fehling, A. (2013, December 25). In Christmas message, Snowden tells Britons "privacy matters" [The Two-Way web log post]. National Public Radio. Retrieved from http://www.npr.org

Garfield, C. (1987). *Peak performers: The new heroes of American business*. New York, NY: William Morrow.

Hareven, T. K. (1991). The history of the family and the complexity of social change. *The American Historical Review, 96*, 95–124. doi:10.2307/2164019

Harmon, A. (2006, April 12). Seeking ancestry in DNA ties uncovered by tests. *The New York Times*. Retrieved from http://www.nytimes.com

Harter, S., (2002). Authenticity. In C. R. Snyder & S. J. Lopez (Eds.), *Handbook of positive psychology* (pp. 382–394). New York, NY: Oxford University Press.

Hernholm, S. (2014, February 21). Authentic self expression [Video]. Available at http://www.youtube.com

Hogenboom, M. (2014, August 22). Study creates "time travel" illusion. British Broadcasting Corporation. Retrieved from http://www.bbc.com

Hookway, N., & James, S. (Eds.). (2015). Authentic [Special issue]. *M/C – A Journal of Media and Culture, 18*(1). Retrieved from http://www.journal.media-culture.org.au

Kaplan, E. (2015, April 26). What role do you want to play? *The New York Times*, p. SR9. Retrieved from http://www.nytimes.com

Kuhn, M. H. (1948). Meet the Amish: A pictorial study of the Amish people. *Social Forces, 27*(2), 174–175.

Luhrmann, T. M. (2014, July 27). Where reason ends and faith begins. *The New York Times*, p. SR9. Retrieved from https://www.nytimes.com

Malin, H. (2011). American identity development and citizenship education: A summary of perspectives and call for new research. *Applied Developmental Science, 15*(2), 111–116.

Marino, G. (2014, May 17). A life beyond "do what you love" [Web log post]. *The New York Times*. Retrieved from http://opinionator.blogs.nytimes.com

Markus, H. R., & Kitayama, S. (1991). Culture and the self: Implications for cognition, emotion, and motivation. *Psychological Review, 98*(2), 224–253.

Maslow, A. H. (1943). A theory of human motivation. *Psychological Review, 50*(4), 370–379.

Moran, S. (2009). Purpose: Giftedness in intrapersonal intelligence. *High Ability Studies, 20*(2), 143–159.

Moran, S. (2014). What "purpose" means to youth: Are there cultures of purpose? *Applied Developmental Science, 18*(3), 1–13.

Olivares, O. J. (2010). Meaning making, uncertainty reduction, and the functions of autobiographical memory: A relational framework. *Review of General Psychology, 14*(3), 204–211.

Perry, F. (2014). "Keeping it real": try-hards, clones and youth discourses of authentic identity from Hollywood to high school. *Journal of Youth Studies*. doi: 10.1080/13676261.2014.1001828

Pew Research Center. (2014, March). Millennials in adulthood: Detached from institutions, networked with friends. Retrieved from Author website: http://www.pewsocialtrends.org

Putnam, R. D. (2007). E pluribus unum: Diversity and community in the twenty-first century (the 2006 Johan Skytte Prize Lecture). *Scandinavian Political Studies, 30*, 137–174.

Seidman, G. (2013). Self-presentation and belonging on Facebook: How personality influences social media use and motivations. *Personality and Individual Differences, 54*, 402–407.

Shin, J., Hwang, H., Cho, E., & McCarthy-Donovan, A. (2014). Current trends in Korean adolescents' social purpose. *Journal of Youth Development, 9*(2), 16–33.

Singer, N. (2015, April 19). Technology that prods you to take action, not just collect data. *The New York Times*, p. BU3. Retrieved from http://www.nytimes.com

Sullivan, W. M. (2011). Interdependence in American society and commitment to the common good. *Applied Developmental Science, 15*(2), 73–78.

Turchi, M., & Scalese, R. (2014, November 17). And the 2014 word of the year is.... Boston.com. Retrieved from http://www.boston.com

Whitehead, C. (2015, April 5). How "you do you" perfectly captures our narcissistic culture. *The New York Times*, p. MM13. Retrieved from https://www.nytimes.com

Wood, A. M., Maltby, J., Baliousis, M., Linley, P. A., & Joseph, S. (2008). The authentic personality: A theoretical and empirical conceptualization and the development of the authenticity scale. *Journal of Counseling Psychology, 55*(3), 385–399.

19
Gender Fluidity

"Sex" reflects biological attributes like chromosomes, hormones, and genitalia. *"Sexual preference"* indicates to whom the person is erotically attracted. *"Gender"* reflects sense of self, social expectations, and role behaviors (World Health Organization, n.d.). Historically, many societies subjected sexual preference and gender to be corollaries of sex, and endorsed only two patterns: men and women. Males (sex) slept with females (sexual preference) and behaved in masculine ways (gender). Same for females, women, and femininity. Separation of duties into *"his"* and *"hers"* organized agricultural – and by extension, urban – societies for millennia (Pedrero, 1999). These social norms were based on practical considerations of division of labor, reproduction, and security, not on how a person experienced one's body or self.

In the US, the struggle to loosen these *"either/or"* stereotypes has proceeded slowly over the last 50 years (Londono, 2015b) but became a focal part of the public conversation in 2015 with recent famous gender transitions (Haberman, 2015; Miller, 2015). Changing norms open a frontier of experiential, identity, and lifestyle options (Stanley, 2015; Tate, Youssef, & Bettergarcia, 2014). Reproductive technologies, surgery, anti-discrimination laws, and household appliances have removed many of the pressures reinforcing binary gender roles (e.g., Calamur, 2014; News from Elsewhere, 2015). Although nontraditional gender choices have been medicalized as gender identity disorder, the empowerment of gender-fluid individuals turns sex/gender incongruities from a *"disorder"* into a *"self-expression"* (Ehrensaft, 2012). *"I am not a category, I am an 'I'"* privileges personal meaning and choice. The Millennial generation, especially, embraces gender as a spectrum (Rivas, 2015).

Gender fluidity could help some individuals feel more authentic, but also confuse people about how we should interact with each other. Former scripts for everyday social interactions may not work anymore (North, 2014b), especially if gender could change daily or hourly (Miller & Spiegel, 2015). How can we develop new ways to signal and symbolize gender so we can understand each other? How should we handle common tasks that formerly were gendered (Apuzzo, 2015; Scelfo, 2015b) in ways that honor diversity, individuality, and safety (Slotnik, 2015), and avoid hurt feelings, misunderstandings, bias or prejudice, and violence (Pasulka, 2015; Samuels, 2015; Sullivan, 2015a, 2015b)?

A Spectrum of Identities
with Sarah Parker

Gender fluidity is an umbrella term to describe possibilities for gender identity beyond the binary "man" or "woman." In 1955, scholars differentiated biological sex from social role expectations of gender (Money & Ehrhardt, 1972). Gender fluidity has become part of our public lexicon recently, especially with several high-profile media stories about celebrities' gender transitions (Lyall & Bernstein, 2015) and gender-fluid TV characters (Deggans, 2015). This media coverage suggests that gender fluidity relates to only some people.

But a gender spectrum suggests that *everyone* is part of this development. Even individuals who identify as the historical categories of "man" and "woman," now called "cisgender" individuals, are simply two options from a more diverse selection of possibilities. If US society is, indeed, at a tipping point for gender fluidity (Pasulka, 2015), what considerations of fairness, benefits and harms, equality, and other ethical dimensions should be addressed?

Let's imagine the situation of universal gender fluidity, where everyone individually chooses. Some may change their biological characteristics surgically. Others focus on expression via clothing or belongings (but stores are no longer separated into men's and women's departments), or physical fitness (sports and gyms are unisex), or social organizations (untethered from designations like "women's groups," "gentlemen's clubs," or "men's only fraternities"), or entertainments (no longer labeled "chick flicks" or "bromances"). With universal gender fluidity, does "transvestite" make sense given anyone could wear any clothing? What happens to "women's work" if jobs are no longer gendered?

A whole society in transition?

Although less discussed, people who abide by historical gender categories could encounter obstacles. For example, as many people become curious about gender fluidity, feminists have chafed how coverage of celebrities who have transitioned rekindles old stereotypes against which feminism has fought for decades, such as sexualized portrayals of women, the idea that women have different brains, and double standards (Burkett, 2015; Garelick, 2015). Biological females living as gendered women may be discounted or curtailed.

Social and linguistic conventions, if mindlessly used, could cause gender stable individuals to feel uncomfortable and perhaps shamed for their lack of social sensitivity or as new options become popular. While trying to make friends, they end up alienating others. Directional cues and signs for where to find things (e.g., clothing in newly de-gendered department stores) may disappear, and they will have to pay closer attention to their movements and actions. Everyone will have to participate in learning new customs.

The road to this possible future also presents obstacles for individuals who are gender fluid (Alarcon, 2015). They may be marginalized as they have been in the past (Fessler, 2015) or lauded as pioneers (North, 2014b). If gender fluidity becomes a new norm, eventually the pioneers would be viewed as unexceptional because everyone could be expected to make deliberate choices about gender, rather than defaulting to what biological sex used to dictate.

Changes for every body

Gender fluidity advocates that individuals are agents of their own bodies (North, 2014b). In a gender-fluid society, sex reassignment surgery may become more popular as gender transition could become more frequent. Perhaps medical technology may advance to make biological transitions quicker and easier. Perhaps sex reassignment becomes expected as a rite-of-passage for all, undergone one or more times during a lifespan. Or sex reassignment could become less common since people may have other ways to experience genders. People who formerly used surgery to "pass" successfully as the "other gender" may no longer need surgery if they would be accepted without it. Or people may become more comfortable with the sex they are born with, even as their genders change regularly. Since evidence suggests these surgeries are less successful than expected, some may consider them unnecessary risks (Batty, 2004).

Who authorizes and pays for treatments? Currently, the military bans medical hormone therapy for transgender soldiers (Samuels, 2015), but prisoners may not be refused treatment (Apuzzo, 2015), and one state covers transgender medical care under Medicaid (Foden-Vencil, 2015). The distinction between sex and gender is most pronounced with healthcare because drugs for some medical ailments have differential effects on female vs. male bodies. Sex or gender designations on government-issued ID cards may be removed altogether. If so, doctors may still need sex information to determine proper treatment. Will identification documentation become more cumbersome (Feeney, 2013) for people whose biological sex and gender identification do not coincide?

Who declares one's gender, and when?

Children have received special attention as some young people have already declared a desire to transition (Morris, 2015; Thompson, 2015). How might parents, doctors, and psychologists address these children's situations? Should there be a minimum age at which persons can decide for themselves, similar to child emancipation criteria? Should parents have responsibilities to inform children interested in transitioning about the uncertainties and challenges (Padawer, 2012)? What language should be used to be conceptually accurate and age appropriate? Supports, such as children's books and videos, are starting to emerge (e.g., Alter, 2015; Londono, 2015a). Perhaps gender exploration eventually becomes encouraged in the same way career exploration is now. Later generations of parents can be more supportive because they also went through gender exploration experiences in their youth. But how should the first generation of parents of transgender youth be trained to handle these conversations when all they may know is binary gender?

Puberty is when the next generation begins to split more clearly into distinct gender groups based on bodily changes. This process can be difficult for gender-fluid children since their bodies are changing but not into the body they want. Transgender children face directions from their parents to live as their biological gender, which can strain family relationships. Should children receive hormone blockers before puberty? If a psychologist diagnoses a child with gender identity disorder, then the child can undergo treatment. Or should children and families wait to see how pubertal hormones affect the children first (Spiegel, 2008)?

If gender fluidity becomes part of the standard rite-of-passage called "sex education," it may become more common. Baby clothes, toys, sports, nursery décor, and similar currently gender-specific childhood conventions may become obsolete. What was formerly conceived of as generally "good behavior" for children may be reconstrued as "gendered behavior" and reconsidered: for example, girls who sit "properly" or boys who "roughhouse." Educational initiatives that differentiate learning styles or performance based on gender may be questioned (How to educate boys, 2015; Porter, 2015). What does a "gender gap" in math or engineering – or any other educational performance indicator – mean in a gender-fluid world?

Putting your "self" out there

The right to express ourselves as we want to be seen is important for many. What currently is only debated for transgender individuals could become commonplace decisions for all. How should someone dress, move one's body, speak? Until recently, there were few socially accepted, institutional options: for example, admission forms at schools (Padawer, 2014; Scelfo, 2015a; Thompson, 2015) or membership forms on social media (Associated Press, 2012; Molloy, 2014; North, 2014b) limited the check boxes to two genders. But in a fully gender-fluid society, a multiple-choice question for gender may become a write-in question since there is a spectrum of options.

One key issue that arises is: What pronoun describes the person best? A gendered pronoun like he or her, or a gender neutral pronoun like ze or zhe? Individuals generally prefer to set the pronoun (Slotnik, 2015; Stanley, 2015) as well as timing of when it should be used (Sullivan, 2015b). Proposed terminology has proliferated greatly (MacFarquhar, 2015; Rosman, 2015; Scelfo, 2015a, 2015b). Universities and dating sites are expanding their options for students and users to self-categorize (North, 2014a; Scelfo, 2015a, 2015b).

Ideally, laws to protect the safety of and to provide options for gender-fluid students (Thompson, 2015) will no longer be needed since gender fluidity will be normalized. However, at present, California passed such a law because even using the restroom can be fraught with danger for transgender students. While colleges have made progress with gender-neutral bathrooms (Bellware, 2014), other public places have yet to follow suit. The lack of access to safe toilets – due to harassment and attacks – has led to infections and other negative health effects (Davis, 2013).

In a fully gender-fluid society, gender-specific school uniforms would be considered silly. But change takes time. So in the meantime, some universities are changing policies to allow students to wear the uniform of whichever gender they prefer, not one assigned based on their sex (Molloy, 2014), and to have a third gender categorization on official documents and a gender-neutral pronoun (Scelfo, 2015a). Traditional "women's colleges" may address their admission and dismissal policies or just allow individual gender-fluid students to navigate the issues independently (Padawer, 2014).

Equal opportunity play

The sports world has been fraught with gender differentiation and sometimes discrimination. Title IX required women's sports options as well as men's at universities. There have been several attempts to start parallel professional sports leagues for women, although most do not gain the public support parallel to men's leagues. Starting in 1900, the Olympics featured women athletes, but only a few until the mid-20th century. Not until 2012 did the Olympics have women competing in all sports (International Olympic Committee, 2014).

But having parallel or even integrated opportunities for men and women is different than gender-fluid opportunities. What would sports and leisure look like in a gender-fluid world? Since sports are, in part, based on the use of body characteristics – such as height for basketball, or weight for wrestling and football, or agility for running or soccer – would biological sex characteristics still be a relevant criterion, in some way, for team selection, setting records, or the like? Would rules shift to accommodate the spectrum of players, such as tackle vs. no tackle versions of a game, or required protective gear? What would happen to locker room joking and behavior if any type of body and any type of gender identity could intermingle in the same space? Would gender be considered irrelevant altogether? Sports marketing may have to be de-sexualized – less about "bros and beer for the big game" and more about gender-neutral "friends and fun" to be more inclusive to different social group compositions. Or perhaps, in a fully gender-fluid world, if inclusion is based more on common interests and less on visual body or dress cues, people may come and go at sports bars as they please. We might have the opportunity to meet a wider variety of interesting people that way.

Further exploration

1. Imagine you are a lawmaker. As more citizens identify as gender fluid, changing their identification often, how would you address issues related to government-issued identification (such as licenses and passports)?
2. Choose one institution that could be impacted by a radical shift in the number of people who identify as gender fluid. What issues and opportunities would it face and why? Pick one of these potential changes: if you were an institutional leader, how would you address it?

References

Alarcon, D. (2015, March 8). Finding the words. *The New York Times*, p. MM38. Retrieved from http://www.nytimes.com

Alter, A. (2015, June 5). Transgender children's books fill a void and break a taboo. *The New York Times*. Retrieved from http://www.nytimes.com

Apuzzo, M. (2015, April 4). Transgender inmate's hormone treatment lawsuit gets Justice Dept. backing. *The New York Times*, p. A1. Retrieved from http://www.nytimes.com

Associated Press. (2012, July 29). Oxford University changes dress code to meet needs of transgender students. *The Guardian*. Retrieved from http://www.theguardian.com

Batty, D. (2004, July 30). Sex changes are not effective, say researchers. *The Guardian*. Retrieved from http://www.theguardian.com

Bellware, K. (2014, July 18). Gender-neutral bathrooms are quietly becoming the new thing at colleges. *Huffington Post*. Retrieved from http://www.huffingtonpost.com

Burkett, E. (2015, June 6). What makes a woman? *The New York Times*. Retrieved from http://www.nytimes.com

Calamur, K. (2014, December 18). U.S. announces protection for transgender workers [The Two-Way web log]. National Public Radio. Retrieved from https://www.npr.org

Davis, M. (2013, June 24). Transgender people need safe restrooms. *Huffpost Gay Voices*. Retrieved from http://www.huffingtonpost.com

Deggans, E. (2015, January 12). Big wins for "Transparent" make it clear: TV's undergoing a revolution [The Two-Way web log]. National Public Radio. Retrieved from https://www.npr.org

Ehrensaft, D. (2012). From gender identity disorder to gender identity creativity: True gender self child therapy. *Journal of Homosexuality, 59*(3), 337–356.

Feeney, N. (2013, July 10). Identity crisis: Changing legal documents no easy task for transgender individuals. *Time*. Retrieved from http://healthland.time.com

Fessler, P. (2015, May 27). Casa Ruy is a "chosen family" for trans people who need a home. National Public Radio. Retrieved from https://www.npr.org

Foden-Vencil, K. (2015, January 10). In Oregon, Medicaid now covers trans medical care [Shots web log post]. National Public Radio. Retrieved from https://www.npr.org

Garelick, R. (2015, June 3). The price of Caitlyn Jenner's heroism. *The New York Times*. Retrieved from http://www.nytimes.com

Haberman, C. (2015, June 14). Beyond Caitlyn Jenner lies a long struggle by transgender people. *The New York Times*. Retrieved from http://www.nytimes.com

How to educate boys [Letters to the Editor]. (2014, February 28). *The New York Times*. Retrieved from http://www.nytimes.com

International Olympic Committee. (2014, May). Factsheet: Women in the Olympic movement. Retrieved from http://www.olympic.org

Londono, E. (2015a, May 8). Families share stories of raising transgender kids. *The New York Times*. Retrieved from http://www.nytimes.com

Londono, E. (2015b, May 18). Increasingly visible, transgender Americans defy stereotypes. *The New York Times*. Retrieved from http://www.nytimes.com

Lyall, S., & Bernstein, J. (2015, February 6). The transition of Bruce Jenner: A shock to some, visible to all. *The New York Times*, p. A1. Retrieved from http://www.nytimes.com

MacFarquhar, L. (2015, January 5). Dos and don'ts: Heavy petting. *The New Yorker*, p. 18.

Miller, C. C. (2015, June 8). The search for the best estimate of the transgender population. *The New York Times*. Retrieved from http://www.nytimes.com

Miller, L., & Spiegel, A. (2015, January 9). Radiolab presents: Invisibilia. Retrieved from http://www.radiolab.org

Molloy, P. M. (2014, February 13). Facebook expands gender options for trans and gender-nonconforming users. *Advocate*. Retrieved from http://www.advocate.com

Money, J., & Ehrhardt, A. A. (1972). *Man and woman, boy and girl: Differentiation and dimorphism of gender identity from conception to maturity*. Baltimore, MD: Johns Hopkins University Press.

Morris, R. (2015, January 12). Transgender 13-year-old Zoey having therapy. British Broadcasting Corporation. Retrieved from https://www.bbc.com

News from Elsewhere. (2015, January 21). Turkey: First shelter for transgender people opens [Web log post]. British Broadcasting Corporation. Retrieved from https://www.bbc.com

North, A. (2014a, December 18). Janet Mock tells the future: Trans people's stories, and safety on Twitter [Web log post]. *The New York Times*. Retrieved from http://op-talk.blogs.nytimes.com

North, A. (2014b, November 19). How OkCupid has become more inclusive on gender and sexuality [Web log post]. *The New York Times*. Retrieved from http://op-talk.blogs.nytimes.com

Padawer, R. (2012, August 8). What's so bad about a boy who wants to wear a dress? *New York Times*. Retrieved from http://www.nytimes.com

Padawer, R. (2014, October 15). When women become men at Wellesley. *The New York Times*. Retrieved from http://www.nytimes.com

Pasulka, N. (2015, January 8). After Leelah Alcorn's suicide, trans youth fight broader bias [Code Switch web log post]. National Public Radio. Retrieved from https://www.npr.org

Pedrero, M. (1999). *Agricultural censuses and gender considerations – concept and methodology*. Rome, Italy: Food and Agricultural Organization of the United Nations. Retrieved from http://www.fao.org

Porter, E. (2015, March 10). Gender gap in education cuts both ways. *The New York Times*. Retrieved from http://www.nytimes.com

Rivas, J. (2015, February 3). Half of young people believe gender isn't limited to male and female. *Fusion*. Retrieved from http://fusion.net

Rosman, K. (2015, June 5). Me, myself and Mx. *The New York Times*. Retrieved from http://www.nytimes.com

Samuels, D. J. (2015, February 23). The next big military civil rights issue: Transgender service [Taking Note web log post]. *The New York Times*. Retrieved from http://takingnote.blogs.nytimes.com

Scelfo, J. (2015a, February 3). A university recognizes a third gender: Neutral. *The New York Times*, p. ED18. Retrieved from https://www.nytimes.com

Scelfo, J. (2015b, February 8). They. *The New York Times*, pp. ED18–20.

Slotnik, D. E. (2015, April 25). In interview, Bruce Jenner says he's transitioning to a woman. *The New York Times*, p. B6. Retrieved from http://www.nytimes.com

Spiegel, A. (2008, May 8). Parents consider treatment to delay son's puberty. National Public Radio. Retrieved from http://www.npr.org

Stanley, A. (2015, April 25). Bruce Jenner, embracing transgender identity, says "It's just who I am." *The New York Times*. Retrieved from http://www.nytimes.com

Sullivan, M. (2015a, February 11). *The Times* and transgender issues (part 1 of 2): The pronouns [Public Editor web log post]. *The New York Times*. Retrieved from http://publiceditor.blogs.nytimes.com

Sullivan, M. (2015b, February 12). *The Times* and transgender issues (part 2 of 2): On Bruce Jenner [Public Editor web log post]. *The New York Times*. Retrieved from http://publiceditor.blogs.nytimes.com

Tate, C. C., Youssef, C. P., & Bettergarcia, J. N. (2014). Integrating the study of transgender spectrum and cisgender experiences of self-categorization from a personality perspective. *Review of General Psychology, 18*(4), 302–312.

Thompson, N. (2015, March 5). Transgender students learn to navigate school halls. Youth Radio, National Public Radio. Retrieved from https://www.npr.org

World Health Organization. (n.d.). What do we mean by "sex" and "gender"? Retrieved August 21, 2015, from http://www.who.int

20
Average as Optimum

Reports in academic journals or mass media state, "On average," an American woman is 5'4", or a student scores 497 on the SAT, or a person can hold seven (plus or minus two) items in working memory (Miller, 1956). These averages – also called means – could help design kitchen cabinet heights, develop a readability program, or write retainable instructions.

But what happens when the "average" is considered the "optimum" – when a typical description becomes a normative ideal? Especially when describing persons, what effect might average traits or behaviors have on how individuals and groups actually behave? In some ways, this use of the average as an ideal exemplifies the classic "is-ought" problem. Descriptive statements segue into normative statements: what is currently occurring is viewed as what ought to occur (Kohlberg, 1971).

Social media, search engine analytics, and wearable gadgets promote the quantification of behaviors and traits into the "quantified self" (Singer, 2015). The Internet provides huge, diverse samples that statistics turn into one number. If we are what can be measured and then statistics collate these measurements into averages, do we understand what averages and statistics represent (Kristof, 2015)? As Big Data promotes the tyranny of the average (Lohr, 2012), will diversity be lost?

We might strive to do what we think these numbers say we <u>ought</u> to do. If all of us do that, then the average is no longer an average of a distribution. It is conformity. Although we often think we want to "stand out from the crowd" and be anything but average, in general, we feel more comfortable when we fit in (Asch, 1948). We tend to imitate others (Bandura, 1977), as well as follow the crowd when we're unsure ("when in Rome, do as the Romans"). With the Internet, "those around us" has become a much larger reference group. As each of us aims to fit in with these numbers, does the "me" eventually become the "mean"?

What Does the "Mean" Really Mean for Me?

with Eliana Hadjiandreou

Variability is everywhere: heights, intelligence, social status, self-confidence. Statistics provide ways to represent variability, including distributions, frequencies, and correlations. In particular, statistics give us a tool to tame variability: the average. This one number is shorthand to capture the "gist" of a phenomenon.

This "average" way of thinking streamlines productivity. For example, instead of garments made for a particular body, they become standardized and "off the rack": individuals conform a size 8 or 10 cut to an "average" height/weight combination. Schools can quickly see which students are excelling and which are lagging behind (Siegler, DeLoache, Eisenberg, & Saffran, 2014).

People prefer to fit in

If the "average" is interpreted as "the majority," the average enters the public consciousness as a representation of "how most people are" rather than "a middle value of a sample of people." This shift in interpretation suggests a *norm* to which a majority adheres, and therefore, that we should follow if we wish to be normal (Bendor & Swistak, 2001; Miller, 1999).

Media reports that generalize "people are doing X" can serve as "nudges" (Thaler & Sunstein, 2009) toward how we are supposed to feel, think, or act. Large-scale psychology experiments done on social media suggest these nudges work, for example, for emotional contagion (Kramer, Guillory, & Hancock, 2014) and voting behavior (Bond, Fariss, Jones, Kramer, et al., 2012). These averages, then, become self-fulfilling prophecies: more people become increasingly like the average behavior over time. Although, initially, the average represented only a midpoint in a distribution, our belief that it is an optimum to which we should aspire turns the average into the majority or, at the extreme, into total conformity.

What would optimizing the average look like?

Consider what everyone's life *should* look like if the average is considered optimal. Collating data from several government and polling sources from 2005 to 2013 (O'Keefe, 2012; Russell, 2011; US Department of

Labor, 2014), each of us is a 38-year-old married woman living with a child and a pet in a mortgaged house in a city in the state we grew up in. Family is very important to us, we are generally friendly to neighbors, and we give to charities. We graduated high school and took some college courses but didn't earn a degree. To contribute to a household income around $50,000, we work 40 hours a week doing administrative work for a private company.

We watch about two hours of television a day. We like to go to the mall. We're overweight by medical standards, but we're generally healthy so we don't worry about it. We have no trouble finding larger-sized clothes because overweight is average. We still have a landline telephone at home, but also a cell phone and Internet access. We have less than $100,000 in savings and don't directly own stock, but there may be some stock in our retirement plan (we don't really know).

Life is not problem-free. Traffic is horrendous because everyone drives everywhere. No one knows how to move forward on pressing social problems because the average education and average workplace are antagonistic to creativity, even though this skill might be nominally promoted. Although we tend not to be proactive toward trending social problems, once a problem becomes mainstream (i.e., average), we try to do our part. For example, we recycle. The average-as-optimal turns the average into the ubiquitous.

The average describes no one

Does this optimally average person describe you? The above depiction of the average American – like all averages – is a fiction. It takes a bird's-eye view on a sample of numbers, but it represents no particular person in real life (although one author searched for the average American; O'Keefe, 2012). When stated in the abstract, an average seems innocuous. But when applied to real life, an average can become absurd. For example, the average American family has 2.3 children (Carlson & Pelletier, 2002). It's disturbing to think what life is like for the 3 child. Or as a columnist humorously pointed out, since America is composed of approximately equal numbers of men and women, the average adult has "one ovary and one testicle" (Kristof, 2015).

More serious examples can be found (Rose, 2013): The average of a sample of pilots' bodies was used to design US Air Force cockpits. Adjustable seats were added quickly because the cockpit was not comfortable or functional for any particular pilot. Fifty thousand intellectually bright students drop out since the average school day doesn't challenge them. Similarly, dosage of medicines and recommended daily allowances of nutrients are based on an average American body, which could result in overdoses for people who weigh less.

The average does not tell us what to do next

The average is a snapshot at a particular point in time in the past because it takes some time to collect data and analyze it to produce the average. So using averages as representations of persons or populations always looks backward. It does not provide clear guidance for what we should do *moving forward*.

If the averages are from polls of the general population, they do not represent truth or expert opinion. They collate amateur observations. Let's say we read that American adults eat, on average, 32 pounds of eggs in a year, should we adjust our own diet? If we find out the average man has 20 sex partners by age 30, and we have less, does that mean we're undesirable? If the average grade point average is 3.3, is it time for me to get my child with the 2.5 GPA a tutor or educational software?

And since averages change as populations are resampled, how much time, effort, and money do we spend keeping up with the average? In the last 50 years, the average American, for example, has grown heavier, older, more educated, and less religious, but they still own a home (Plaue, 2012). So if we understand the average as the optimum, and a new average is published every two years, we may feel "yanked around" about what we should be doing.

Is the average short-sighted?

Turning the average into the optimum effectively makes invisible the "better end" of a distribution. If we say that everyone should reach for the average, this aim may be helpful to obese people, people with high blood pressure, sedentary people, and illiterate people. But this focus provides no guidance for – nor even recognition of – underweight people, people with low blood pressure, super-athletes, and voracious readers. The assumption is that these individuals are "fine on their own." But what may be the case is that they are less researched so less is known about them. As the 1990s focus on anorexia and the 2010s examination of excessive exercise suggest (see Reynolds, 2012, 2015), there could be "too much of a good thing."

Furthermore, since "above average" is considered being "better" on some trait, and if average equals optimum, then above-average individuals should "tone down" whatever we do well. If we are intellectually gifted, or talented musicians, or Olympic athletes, or exceptionally charismatic, then trying to fit in means falling below our potential. Although we might like to think that such perspectives don't occur, consider US education today and its emphasis to close "gaps" between low achievers and the average. The average often falls between "basic" and "proficient" scores (see Child Trends Data Bank, 2014, p. 3), which isn't an ambitiously high bar – it's average. Students who score "above average" or "advanced" are helpful for keeping a school's average high – and off the dreaded "underperforming" list. But these students are a nearly forgotten group (Pandina Scot, Callahan & Urquhart, 2009). Our drive to reach our potential – as individuals and as societies – might stagnate over time.

What if the average became institutionalized and required?

What if "average is the optimum" became so powerful that institutions *mandate* the average? What if variability is significantly reduced or even eliminated? The extreme might be, using recent biological and medical innovations, that the average American is cloned. There would be no more classes, or minorities, or geniuses because everyone is average. Production costs could be minimized because only one average version of cars, cereals, computers, and other products would be produced. There would be no need for polling because, if everyone is average on everything, researchers only need to ask one person. "Average" would become a meaningless word, unnecessary because there is no sample or population.

What adjectives would you use to describe this imaginary scenario? What would it feel like to live in this world if it became real? Of course, this scenario is oversimplified. Even clones don't necessarily have the same experiences, so variability could arise through different perspectives. In a world of interactions, it is difficult to maintain complete conformity – although, over history, totalitarian regimes have tried.

Without diversity, what is possible?

A backlash against average may arise. Despite our leanings toward belonging, a fear of being ordinary, dull, or mediocre also exists (O'Keefe, 2012). We want people to come to agreement, yet we want individual voices to be part of the symphony of democracy. We anchor to the average, but believe we and our children are above average (the Lake

Wobegon Effect; Keillor, n.d.). The self-esteem movement turns all of us into someone "special." The fragmentation of society into a plethora of lifestyles keeps us from comparing ourselves to random others so we do not become unhappy (Quartz & Asp, 2015). Each of us is "unique" and, by using new gadgets, we can continue to differentiate ourselves (Friedman, 2014).

Variation makes families, communities, societies more robust and resilient (Page, 2007). If everyone is not the same, and the group is challenged, there are more possibilities from which to choose a helpful response. If we have differences, we can cooperate and produce something better than each could do separately, or we can catalyze each other's abilities.

Variation puts and keeps the world in motion. If we are all average, there is no impetus to generate momentum. Novel ideas are less likely generated at the average or center, and are more likely found at the edges where the *un*usual is more likely (Fish, 2008). "Groupthink," stemming from too many average-thinking people, tends to be anathema to problem solving. It would be difficult to find interests. Interest is aroused when something is mildly surprising and unexpected – that is, not normal. In a uniformly average world, everything is normal.

The average is a single point that tells us where the middle of a sample's values lie. Statistical information published without consideration of how people follow norms, imitate others, and spread ideas and behaviors may be counterproductive (Postrel, 2005). The result may conform people to expectations derived from a statistical sample, rather than explain real people. If we aim for average and build institutions on a foundation of the average – no matter how large a sample the average is calculated from – are we fostering stagnation rather than potential? Life is lived on the edge of possibility, perpetuated by differentials and gradients. Using averages to solve problems may feel like success. But by also appreciating variability, we glimpse the "adjacent possible" (Johnson, 2010), where innovation lies.

Further exploration

1. What are the pros and cons of familiarity and commonality? What are the pros and cons of variability and diversity?
2. How might we overcome the urge to conform to the average?
3. What effects might reduced variability among people have on society in 50 years? 100 years? 500 years?

References

Asch S. E. (1948). The doctrine of suggestion, prestige, and imitation in social psychology. *Psychological Review, 55*, 250–276.

Bandura, A. (1977). *Social learning theory.* Englewood Cliffs, NJ: Prentice Hall.

Bendor, J., & Swistak, P. (2001). The evolution of norms. *American Journal of Sociology, 6*, 1493–1545.

Bond, R. M., Fariss, C. J., Jones, J. J., Kramer, A. D. I., Marlow, C., Settle, J. E., & Fowler, J. H. (2012, September 13). A 61-million-person experiment in social influence and political mobilization. *Nature, 489*, 295–298. doi:10.1031/nature11421

Carlson, G., & Pelletier, F. J. (2002). The average American has 2.3 children. *Journal of Semantics, 19*, 73–104.

Child Trends Data Bank. (2014, September). *Reading proficiency: Indicators on children and youth.* Retrieved from http://www.childtrends.org/wp-content/uploads/2013/01/29_Reading_Proficiency.pdf

Fish, S. (2008, June 1). Norms and deviations: Who's to say? [Web log] *The New York Times.* Retrieved from http://opinionator.blogs.nytimes.com

Friedman, T. L. (2014, May 20). Four words going bye-bye. *The New York Times.* Retrieved from https://www.nytimes.com

Johnson, S. (2010). *Where good ideas come from.* New York, NY: Penguin.

Keillor, G. (Producer). (n.d.). *The prairie home companion* [Radio show & podcast]. Available from http://prairiehome.org

Kohlberg, L. (1971). From is to ought: How to commit the naturalistic fallacy and get away with it in the study of moral development. In T. Mischel (Ed.), *Cognitive development and psychology* (pp. 151–235). New York, NY: Academic Press.

Kramer, A. D. I., Guillory, J. E., & Hancock, J. T. (2014, June 17). Experimental evidence of massive-scale emotional contagion through social networks. *PNAS, 111*, 8788–8790. doi: 10.1073/pnas.1320040111

Kristof, N. (2015, April 26). Are you smarter than an 8th grader? *The New York Times*, p. SR11. Retrieved from https://www.nytimes.com

Lohr, S. (2012, February 11). The age of big data. *The New York Times*, p. SR1. Retrieved from https://www.nytimes.com

Miller, D. T. (1999, December). The norm of self-interest. *American Psychologist, 54*(12), 1053–1060.

Miller, G. A. (1956). The magical number seven, plus or minus two: Some limits on our capacity for processing information. *Psychological Review, 63*(2), 81–97.

O'Keefe, K. (2012). *The average American: The extraordinary search for the nation's most ordinary citizen, 2nd ed.* New York, NY: Chilva.

Page, S. (2007). *The difference: How the power of diversity creates better groups, firms, schools, and societies.* Princeton, NJ: Princeton University Press.

Pandina Scot, T., Callahan, C. M., & Urquhart, J. (2009). Paint-by-number teachers and cookie-cutter students: The unintended effects of high-stakes testing on the education of gifted students. *Roeper Review, 31*, 40–52. doi: 10.1080/02783190802527364

Plaue, N. (2012, July 7). How the average American has changed since the 1960s. *Business Insider.* Retrieved from www.businessinsider.com

Postrel, V. (2005, August 11). Adding social norms to the usual methodology mix. *The New York Times.* Retrieved from https://www.nytimes.com

Quartz, S., & Asp, A. (2015, April 12). Unequal, yet happy. *The New York Times*, p. SR4. Retrieved from http://www.nytimes.com

Reynolds, G. (2012, September 19). For weight loss, less exercise may be more. *The New York Times*. Retrieved from https://www.nytimes.com

Reynolds, G. (2015, May 6). An unexpected death rattles the fitness community. *The New York Times*. Retrieved from https://www.nytimes.com

Rose, T. (2013, June 19). *The myth of average* [Video]. Retrieved from http://tedxtalks.ted.com/

Russell, C. (2011, July 19). 50 facts about the average American. *American Consumers Newsletter*. Retrieved from www.newstrategist.com

Siegler, R., DeLoache, J., Eisenberg, N., & Saffran, J. (2014). *How children develop*, 4th ed. New York, NY: Worth Publishers.

Singer, N. (2015, April 19). Technology that prods you to take action, not just collect data. *The New York Times*, p. BU3. Retrieved from http://www.nytimes.com

Thaler, R. H., & Sunstein, C. R. (2009). *Nudge: Improving decisions about health, wealth and happiness*. New York, NY: Penguin.

US Department of Labor, Bureau of Labor Statistics. (2014, June 18). *American time use survey – 2013 results* (Publication USDL-14–1137). Retrieved from http://www.bls.gov/news.release/pdf/atus.pdf

21
Big Data

Companies and governments have collected and stored data about us for decades but data were isolated in separate locations. Now data can be aggregated via the Internet (Sangani, 2013). Plus new sensors, keystroke capture, Internet cookies, cameras, e-commerce, and social media can track nearly our every move (Lohr, 2012). Phones, cars, appliances, and fashion accessories can monitor our location, habits, and health.

Our data may be used to personalize services (Goel, 2014), seek opportunities (Lohr, 2012), build communities (Oboler, Welsh, & Cruz, 2012), scout high achievers (Blum, 2015), capitalize on events in real time (Hardy, 2014b), prevent suffering (Bookman, 2015; Greenfieldboyce, 2014), manipulate our actions (Streitfeld, 2015), and even humiliate or blackmail us (Maslin, 2014).

The more we share, the more attention and personalization we can garner, yet the more vulnerable we may become to uses of our data we did not anticipate and may not want (Charles, 2014; Lieber, 2014). Potential costs include decreased privacy, constrained choices, increased biases, job losses, and identities reduced to numerical profiles (Hardy, 2014a; Siegel & Cornish, 2015). As data generators, we face issues of control and identity: who might access our data and misrepresent us? As data users, we face issues of responsibility and transparency: how do we consider repercussions of the knowledge we gain from others' data?

What Does It All Add Up To?

with Michael Masters

Meet Juan. His emails, tweets, posts, purchases, page views, and clicks leave clues online about his lifestyle, preferences, and interests. These clues tailor information provided to him as a data user. The more Juan seeks his dream career, the more job-relevant the ads he receives. Juan's opinions about national border disputes on social media garner him news feeds and filtered posts reinforcing his beliefs. Juan's political comments on news sites lead to automatic campaign updates. Apps that note he loves grandma's tamales send him recipes and coupons for ingredients. Political parties may combine these clues to promote their candidates in relation to job prospects, immigration reform, and family values.

Juan, meet Big Data. Big Data collects and aggregates the unstructured data Juan provides, which can be used for multiple purposes often not specified before his data are captured (Sangani, 2013). Although vaguely anticipated since the 1940s (Press, 2013), Big Data is heralded as the "economic asset" of the 21st century (Lohr, 2012). Recent analytic tools transform these data from patterns into insights and decisions (Lohr, 2013; Manyika, Chui, Brown, Bughin, et al., 2011). Leaders jockey for advantages of Big Data's campaigning (e.g., Cohen, 2013), economic (BBC Staff, 2014), and security (Risen & Lichtblau, 2013) benefits. But controversy over surreptitious collection remains high (Shane, 2015).

Aggregated patterns don't predict individual pathways

As individuals, we both generate and use Big Data. However, we rely on the intermediaries of "data scientists," statisticians, modelers, and programmers to organize and mine these vast databases (Lohr, 2012, 2013, 2014). The risk is that erroneous patterns may result because correlations are more likely by chance as dataset sizes increase (Eisenger, 2015). These errors may lead to wasting resources on poorly conceived services (Kraft, 2015) or, more egregiously, to discrimination against individuals who don't fit the models well (Lohr, 2013).

Data can produce findings that look precise, but these numbers need interpretation (Shaw, 2014): what does the finding mean in real life? Furthermore, correlations only tell us, *on average*, that when one thing occurs, something else is more or less likely to occur. Correlations do not tell us what caused something, nor predict what will happen, nor decide individual cases – even though this information is what we really want. Big Data might produce a graph of the rise and fall of "cool" as a slang term (Grimes, 2013), or decipher patterns of our past good deeds (Carey, 2014), or discover drug interactions from medical records (Greenfieldboyce, 2014). Big Data fails to tell us *why* "cool" is popular again, whether we will be kind *today*, or if we as *individuals* will suffer an adverse drug reaction. Although we may hope that genomic databases lead to "precision medicine" tailored to our specific bodies (Harris, 2015), the findings may not apply to us if our genomes don't fit the targeted profile.

Measurement does not equal meaning: we still must decide how to proceed in our own lives (Peysakhovich & Stephens-Davidowitz, 2015). We could collect data about ourselves for self-improvement (Lohr, 2015a). However, generating data we can't make sense of could cause anxiety or harm from unnecessary treatments (Ornstein, 2015; Singer, 2015b; Wayne, 2015). Comparing our data to others' data may motivate us (Westervelt & Kamenetz, 2015) or make us obsess (Rich, 2015).

Lost in the crowd, or caught in a fishbowl?

Combining everyone's data – even without names or locations – may not hide our identities (Singer, 2015a). We worry about privacy: others may discover something about us we don't want known (Lohr, 2013), such as our dating habits (Stephens-Davidowitz, 2015), especially if portrayals are intentionally distorted (Singer, 2014). We don't enjoy helplessness regarding disclosure, yet we accept it as the price for services we depend on (Madden, 2014; Siner, 2014).

Few of us proactively secure our data, even as revelations suggest that governments' and companies' economic interests overshadow their ethical responsibilities (Bilton, 2014; Shane, 2015). However, new tools may at least let us see who use our data (Lohr, 2014), although such transparency may not lead to more control over our own data (Eisenger, 2015) even if we buy costly other gadgets to protect ourselves (Angwin, 2014).

As we come to believe that data collectors do not have our best interests at heart, we may try to disguise ourselves. Returning to Juan, he is interested in Marxism, once cheated on a test, and recently broke up with his girlfriend. But he avoids these topics or intentionally put "red herrings" in his posts for fear they might stain his reputation. This creates a "garbage in garbage out" problem for Big Data because misleading data leads to misleading conclusions. We may post only what we think is acceptable or garners rewards (Jastrzebski, 2015). But such deceit can make Big Data less useful for improving services, especially services that rely on user reviews and ratings (Greenfield, 2012). Furthermore, lying may backfire as we forget the inherent value of the truth, come to believe our own lies, and lose our authenticity.

We want to protect ourselves from those who might use our data to make us miserable (Siner, 2014). Big Data describes old crimes in new language: stealing becomes hacking, impersonation becomes identity theft, hijacking becomes Trojan horses, and humiliation becomes leaks (e.g., Reuters, 2015). Although we may not know the cybercriminals or cyberbullies, the attacks feel personal because they know so much about us.

Do we count?

On one hand, those of us with an online presence enjoy websites catering to us. But this feedback may amplify what we already believe, resulting in an "echo chamber." We interact with like-minded people or read confirming information. We may feel comfortable, but when we are funneled only ideas statistically related to our current ways of thinking, we miss valuable surprises that jolt us out of complacency. Despite the breadth of data aggregated, Big Data could become Narrow Data and reinforce our existing stereotypes.

On the other hand, those who live "off the grid" are not included in analyses, skewing results. Most Big Data contain Internet-savvy WEIRD samples (western, educated, industrialized, rich, and democratic; Heinrich, Heine, & Norenzayan, 2010). Those in isolated or poor parts of the world, without Internet access, are left out. They don't have privacy concerns because they are digitally invisible. Even within wealthier countries, what happens to people who are not considered good "monetizer prospects" for Internet companies? Is an online version of "one percenters" arising who skew society based on data rather than money?

What if everyone knows everything about each other?

An extreme scenario of our sharing society is that no one privately owns data. All data are public and transparent. We might no longer create personal memories because we rely on our digitized collective memory – after all, many of us offload remembering phone numbers to our smartphones. Through Big Data, we might experience others' memories vicariously, or we might re-experience events from a more objective perspective. Would that make our identities equal to the bits stored on computer servers? Would amnesia be cured with a download?

If these data are not secure, then our memory may be lost by deletion, hacking, or alteration – and without individual memories, no one would know any better. Even if these data are secure, potential issues arise. Would trust become ubiquitous because Big Data is transparent, or irrelevant because Big Data is omniscient? Would relationships become easier because we could quickly discern what we have in common, or boring because we no longer need to disclose through conversation (Turkle, 2011)? What happens to data no one wants to know or remember – tragedies, traumas, humiliations, or hostilities? Could we ever live down our past mistakes?

Could Big Data itself take over?

Another extreme scenario is that Big Data analysis makes the big decisions in high-stakes situations like medicine, finance, and policing (Tufekci, 2015). Sophisticated analysis tools – like artificial intelligence – can learn patterns on their own. Eventually, these tools may need less human direction. Civil rights advocates call for "algorithmic accountability" to make sure that decisions are based not only on numbers-crunching but also on meaning-making through human creativity, multiple perspectives, and nuance (Lohr, 2015b).

Perhaps the most extreme scenario is that Big Data centralizes power in itself. As algorithms hone their "intelligence" through continued feeding on more data, perhaps *they* become *our* masters (Brooks, 2014; Hardy, 2014a; Wall, 2014). They may keep a few engineers to provide software tune-ups, but maybe they become relatively self-sufficient and even more human-like.

Returning to Juan: In a society increasingly mediated by gadgets and "cloud" software, he – and we – face quandaries: Will the economic and social benefits for us as data users overwhelm the ethical concerns we have as data generators? Is it enough to know *what* (data), or is it important to understand *why* and *how* (meaning)? What is the proper proportion or integration of quantified data and qualified wisdom?

Further exploration

1. How might Big Data affect people who do not use their services at all?
2. Argue for or against the following claim: Big Data will make human foresight and creativity less necessary.
3. If everything that everybody knows is part of Big Data, what might happen to wonder?

References

Angwin, J. (2014, March 3). Has privacy become a luxury good? *The New York Times*. Retrieved from https://www.nytimes.com

BBC Staff. (2014, March 19). Alan Turing Institute to be set up to research big data. British Broadcasting Corporation. Retrieved from https://www.bbc.com

Bilton, N. (2014, November 27). Moral issues bedevil Silicon Valley. *The New York Times*, p. E2. Retrieved from http://www.nytimes.com

Blum, R. (2015, March 23). The big shift: Infields spin in response to data explosion. Associated Press. Retrieved from http://m.apnews.com

Bookman, T. (2015, June 8). Insurer uses personal data to predict who will get sick [Shots web log post]. National Public Radio. Retrieved from http://www.npr.org

Brooks, D. (2014, October 30). Our machine masters. *The New York Times*, p. A31. Retrieved from http://www.nytimes.com

Carey, B. (2014, September 11). In a study, text messages add up to a balance sheet of everyday morality. *The New York Times*, p. A19. Retrieved from https://www.nytimes.com

Charles, D. (2014, January 22). Should farmers give John Deere and Monsanto their data? [The Salt web log post]. National Public Radio. Retrieved from https://www.npr.org

Cohen, M. (2013, June 19). From campaign war room to big-data broom [Web log post]. *The New York Times*. Retrieved from http://bits.blogs.nytimes.com

Eisenger, J. (2015, February 12). In an era of disclosure, an excess of sunshine but a paucity of rules [Dealbook web log post]. *The New York Times*, p. B5. Retrieved from http://dealbook.nytimes.com

Goel, V. (2014, August 12). As data overflows online, researchers grapple with ethics. *The New York Times*. Retrieved from https://www.nytimes.com

Greenfield, R. (2012, April 25). How to protect yourself from Yelp's robot protection. *The Atlantic Wire*. Retrieved from http://www.thewire.com

Greenfieldboyce, N. (2014, July 21). Big Data peeps at your medical records to find drug problems [Shots web log post]. National Public Radio. Retrieved from https://www.npr.org

Grimes, W. (2013, December 24). Big data becomes a mirror [Review of "Uncharted" by Erez Aiden and Jean-Baptiste Michel]. *The New York Times*. Retrieved from http://www.nytimes.com

Hardy, Q. (2014a, November 23). What could computing mean to your job [Web log post]. *The New York Times*. Retrieved from http://bits.blogs.nytimes.com

Hardy, Q. (2014b, December 11). Cisco's networks will analyze us [Web log post]. *The New York Times*. Retrieved from http://bits.blogs.nytimes.com

Harris, R. (2015, January 30). Obama wants funding for research on more precise health care [Shots web log post]. National Public Radio. Retrieved from https://www.npr.com

Heinrich, J., Heine, J., & Norenzayan, A. (2010, July). Most people are not WEIRD. *Nature, 466*(1), 29.

Jastrzebski, S. (2015, April 20). Social media can help track tornadoes, but was that tweet real? National Public Radio. Retrieved from http://www.npr.org

Kraft, D. (2015, March 29). Can data stop car wrecks? *The New York Times*, p. SR10. Retrieved from http://www.nytimes.com

Lieber, R. (2014, August 15). Lower your car insurance bill, at the price of some privacy. *The New York Times.* Retrieved from https://www.nytimes.com

Lohr, S. (2012, February 11). The age of big data. *The New York Times.* Retrieved from http://www.nytimes.com

Lohr, S. (2013, June 19). Sizing up big data, broadening beyond the internet [Web log post]. *The New York Times.* Retrieved from http://bits.blogs.nytimes.com

Lohr, S. (2014, August 18). Xray: A new tool for tracking the use of personal data on the web [Web log post]. *The New York Times.* Retrieved from https://www.nytimes.com

Lohr, S. (2015a, April 1). Healed by his own data. *The New York Times,* p. B1. Retrieved from http://www.nytimes.com

Lohr, S. (2015b, April 7). Maintaining a human touch as the algorithms get to work. *The New York Times,* p. A3. Retrieved from http://www.nytimes.com

Madden, M. (2014, November 12). Public perceptions of privacy and security in the post-Snowden era. Pew Research Center. Retrieved from http://www.pewinternet.org

Manyika, J., Chui, M., Brown, B., Bughin, J., Dobbs, R., Roxburgh, C., & Byers, A. (2011). Big Data: The next frontier for innovation, competition, and productivity. McKinsey Global Institute. Retrieved from http://www.mckinsey.com

Maslin, J. (2014, August 27). Leaving money and privacy on the table [Review of "What stays in Vegas" by Adam Tanner]. *The New York Times.* Retrieved from http://www.nytimes.com

Oboler, A., Welsh, K. & Cruz, L. (2012). The danger of Big Data: Social media as computational social science. *First Monday, 17*(7), 1–17. Retrieved from http://firstmonday.org

Ornstein, C. (2015, April 6). Tracking your own health data too closely can make you sick [Shots web log post]. National Public Radio. Retrieved from https://www.npr.org

Peysakhovich, A., & Stephens-Davidowitz, S. (2015, May 2). How not to drown in numbers. *The New York Times.* Retrieved from http://www.nytimes.com

Press, G. (2013, May 9). A very short history of Big Data. *Forbes,* 1–2. Retrieved from http://www.forbes.com

Reuters. (2015, June 15). Sex, lies and debt potentially exposed by U.S. data hack. *The New York Times.* Retrieved from https://www.nytimes.com

Rich, M. (2015, May 11). Some schools embrace demands for education data. *The New York Times.* Retrieved from http://www.nytimes.com

Risen, J., & Lichtblau, E. (2013, June 8). How the U.S. uses technology to mine more data more quickly. *The New York Times.* Retrieved from https://www.nytimes.com

Sangani, P. (2013, October 18). My book "Big Data" represents benefits and possibilities in the future: Viktor Mayer-Schonberger. *The Economic Times.* Retrieved from http://articles.economictimes.indiatimes.com/

Shane, S. (2015, May 19). Snowden sees some victories, from a distance. *The New York Times.* Retrieved from https://www.nytimes.com

Shaw, J. (2014, March/April). Why "big data" are a big deal. *Harvard Magazine,* pp. 30–35, 74–75. Retrieved from https://harvardmagazine.com

Siegel, R., & Cornish, A. (2015, March 23). Robot reporters: Software turns raw data into sports, financial reports. National Public Radio. Retrieved from http://www.npr.org

Siner, E. (2014, March 11). The internet will be everywhere in 2025, for better or worse [All Tech Considered web log post]. National Public Radio. Retrieved from http://www.npr.org

Singer, N. (2014, September 7). OkCupid's unblushing analyst of attraction. *The New York Times*, p. SR2. Retrieved from https://www.nytimes.com

Singer, N. (2015a, January 29). With a few bits of data, researchers identify "anonymous" people [Web log post]. *The New York Times*. Retrieved from http://bits.blogs.nytimes.com

Singer, N. (2015b, April 14). Report questions whether health apps benefit healthy people [Web log post]. *The New York Times*. Retrieved from http://bits.blogs.nytimes.com

Stephens-Davidowitz, S. (2015, January 24). Searching for sex. *The New York Times*. Retrieved from http://www.nytimes.com

Streitfeld, D. (2015, January 30). Ratings now cut both ways, so don't sass your Uber driver. *The New York Times*, p. A1. Retrieved from http://www.nytimes.com

Tufekci, Z. (2015, April 19). The machines are coming. *The New York Times*, p. SR4. Retrieved from http://www.nytimes.com

Turkle, S. (2011). *Alone together: Why we expect more from technology and less from each other*. New York, NY: Basic Books.

Wall, M. (2014, October 9). Could a big data-crunching machine be your boss one day? British Broadcasting Corporation. Retrieved from https://www.bbc.co.uk

Wayne, T. (2015, January 2). The unending anxiety of an ICYMI world. *The New York Times*. Retrieved from http://www.nytimes.com

Westervelt, E., & Kamenetz, A. (2015, March 15). Six things we learned at South by Southwest EDU. National Public Radio. Retrieved from http://www.npr.org

Part VI
Social Connectors

22
The Right to Be Forgotten

What ethical relationships should we forge among freedom of personal expression, the public's right to know, privacy, and control of information about us? That is the crux of the right to be forgotten. What makes the right to be forgotten creative is that, for most of history, people have struggled to be remembered. To create a legal right to remove oneself from the public record is rather revolutionary. It is also an idea that captures the current zeitgeist – or *"spirit of the times."* In a sense, this idea encapsulates Andy Warhol's quip about everyone being briefly famous, except the fame may not be for something we want to be famous for, and the afterlife of our fame may continue indefinitely.

The right to be forgotten is a European legal concept (Rosen, 2012) that may gain traction in the United States, especially as privacy concerns continue to grow (Scott, 2014). The idea stems from European laws and court decisions that give residents control over online information about themselves (Garsd, 2015). Europeans can request that search engines remove specified links to information about them that can be found by using their names. Embarrassing photos, dismissed legal skirmishes, or an irrelevant fact can be made less accessible (Toobin, 2014).

On one hand, we'd like past stupid mistakes to not follow us forever. On the other hand, memories and records underlie personal identity, decision-making capabilities, trusting relationships, and historical accounts. A world with the right to be forgotten as the norm may manifest what philosopher George Santayana warned, "Those who cannot remember the past are condemned to repeat it."

Disappearing Acts
with Christopher Charles Canieso

Although data protection laws have existed in Europe for several years, a landmark court decision in 2014 required Internet search engines to establish procedures for Europeans to request removal of unwanted links to information about themselves (Ball, 2014). Searches on a person's name would no longer find that information through European sites (Shahani, 2014).

To make a link disappear, we must file a claim only about our own information, provide credible explanations, and allow publishers to respond (Rawlinson, 2015). Just because the content is unwanted doesn't mean it will be removed (Essers, 2015).

"Court *is* now in session..."

What is *not* novel is removal of content or links. Websites remove libelous, copyrighted, violent, pornographic, and unlawful content (Toobin, 2014). What *is* new is that search engines have become the judges of what to remove. Although the intent of the data protection laws was, in part, to reduce the power of search engines, these laws may result in a "revenge effect" (Tenner, 1996) of increasing their power via adjudication of requests (Toobin, 2014).

Governments still supervise – and so far, have agreed with – most search engines' decisions (Rawlinson, 2015). Furthermore, internet companies seem uncomfortable with this regulatory role (Rawlinson, 2015), despite their burgeoning staffs to evaluate requests (Essers, 2015). As they develop expertise and protocols for these judgments, might they find a way to use to their advantage the requests as sources of information about what bothers people, thus further strengthening their algorithms to "monetize" our emotions and life events? And *should* the search engines hold the responsibility to adjudicate each European country's balance of privacy and public interest?

Who *is* responsible for our data?

If the online companies that collect data are responsible, then to whom are they accountable? Many afford almost all *rights* to themselves yet give users the *responsibility* to opt out. Perhaps board members, executives, managers, and employees of these companies should implement all corporate decisions regarding personal data on themselves.

If the government is responsible, which government(s)? European leaders want to extend the reach of the "right to be forgotten" beyond the European Union (Scott, 2015a), but cultures disagree on focus: Americans worry more about government abuse, whereas Europeans focus on corporate abuse (Scott, 2014). Many governments face public hostility for their own large-scale data tracking (Erlanger, 2013; Sayare, 2013; Smale, 2015; Steinhauer & Weisman, 2015). Digital protection is a hot political issue (Shahani, 2015). But corporations and users span several jurisdictions (Scott, 2015b). Will one country's rules prevail, or will the Internet break into national fiefdoms (Toobin, 2014)?

If users are responsible, then why do we post information that makes us vulnerable to ridicule or culpability? Sharing technology makes it difficult to control information because others post photos of us, or quote us, or otherwise pass along information. What about child users? Several bills have proposed an "eraser button" for content children post online, but these efforts tend to stall (Scott, 2014).

If responsibility is shared, then how is it allocated across users, corporations, and governments? What would be the ethical foundation of the responsibilities: actual or potential harm, existing laws like copyright, rights related to free speech and access to information, or opt-out or opt-in mechanisms (Scott, 2014)?

Mirror, mirror

Memories make us who we are. Scenarios when memory fails (e.g., Golin & Bregman, 2004; Todd & Todd, 2001) show people their losing sense of self, goals, reasons, and relationships. Online selves differ from our everyday selves because they are mediated through websites and because they can live on as long as the servers have electricity to stay on, even after our deaths (Biersdorfer, 2015). Could the right to be forgotten stimulate new forms of identity? If we develop an online identity, and part of that identity is removed, could it alter who we are offline?

Online identities are easily editable. Indeed, businesses that monitor and correct deviations from desired online reputations are in growing demand (Garsd, 2015). We can present false self-images easily, but as long as other sources of information remains available, misrepresentations can be discovered (Kauffmann, 2014). If the right to be forgotten removes information that conflicts with our online identity, it becomes more difficult to get to know one another. The right to be forgotten can shift the dynamic of deception because, if information is removed, the ratio of false-to-true information may rise, and deception becomes harder to detect (Bond & DePaulo, 2006).

Furthermore, we may come to deceive ourselves that the online identity we've curated is who we really are (Goleman, 1996). If the right to be forgotten removes disconfirming evidence about our ideal self, then we lack information to correct our misperceptions. And curating an ideal self-image online does not mean we behave like the self we portray (Tavris & Aronson, 2007). We can become increasingly disconnected from our identities.

This situation may reflect an online version of Goffman's (1959) drama of life: we perform for each other; we make others believe our persona is true and trustworthy. What is new is how we are now expected to condense our selves into short posts, and update our selves relentlessly. As we feel pressured to present only our ideal self, we may expect each other to live up to our curated ideal selves. When we don't, we reject each other. Our Goffmaniacal self-presentation may take revenge in the inability for us to recognize and accept each other's imperfections. Patience, understanding, empathy, and sympathy may diminish.

Time heals all?

The right to be forgotten may be beneficial by giving our information an expiration date. We live actual events and may share descriptions or pictures of these events via social media. Others may re-send that information to others. Redundancy can create a reliable and accessible record of the event regardless of our wishes to control the information. Searches that used to take considerable effort can now be done in minutes.

Individuals with black marks in their personal histories, such as criminal convictions, may face tough challenges to become socially acceptable again (Toobin, 2014). The United States has a statute of limitations for most crimes. Does the right to be forgotten institute a statute of limitations when past bad acts can be publicly accessible? Although the right to be forgotten does not erase information, it provides a "speed bump" so that searches don't as easily find problematic information about a person; people have to deliberately seek particular information to find it (Mayer-Schonberger, 2009). This required extra effort may give some people a second chance at redemption.

Yet, the Internet's perpetual records may cause trouble even for law-abiding citizens. Social media builds a new form of "personal archives" that post memoir chapters in real time (Good, 2012). Although such records existed before, such as in scrapbooks, now online they can be mined for information, which can be shared more widely. Currently, the uses of information focus on marketing (Opsahl & Reitman, 2013). But those who oppose this data aggregation recall times when such collection started innocently enough, yet turned into nightmares as the information was used to destroyed community welfare (Mayer-Schonberger, 2009). They warn that even if our information is now in trustworthy hands, it is difficult to foresee how the data may be used or misused 10 years hence.

However, using the right to be forgotten to remove the technological links to problematic data may focus more attention on what we want forgotten. For example, after the right to be forgotten lawsuit was won, publicity increased public knowledge of the situation the plaintiff wanted forgotten (Ball, 2014). Similar revenge effects have befallen others who sued for privacy infringement (Parkinson, 2014). Already, the right to be forgotten has spawned websites that track the links that petitions erase (Rawlinson, 2015), further emphasizing what the petitioner didn't want others to know. Not only is the petitioner's embarrassing information still available, it is now flagged as embarrassing.

Who can we trust?

If our curated selves are not trustworthy, and information we search for is incomplete, how do we build trusting relationships with each other? Friendships may become difficult because we may raise our emotional guard. Without honesty and transparency, the appeal of relationships and belonging could diminish. Skepticism can spiral into a lack of empathy and social effort.

History – the aggregation of past events – is regarded as important in cultures worldwide. The right to be forgotten belittles history by erasing personal links that underlie the ability to aggregate information accurately. What is forgotten leaves holes in the historic record, which can influence what history can be written. Could it become possible for an entire country to claim the right to be forgotten? Or for an entire historic event to be erased? Could the claims for online erasure expand to claims for complete erasure in all media, such that, for example, a high school history textbook is missing the chapter on a particular war? When does the right to be forgotten become censorship?

The right to be forgotten could be a boon for criminals, pedophiles, quack doctors, poor workers, deposed dictators, and others who have been previously "called out" for unacceptable behavior. Invoking this right makes the trail of their bad deeds harder to follow. Wealthy and stealthy individuals could cleanse their reputations regularly. Perhaps, the concept of reputation becomes meaningless. The institutional mechanisms to vet each other's character could vaporize. Perhaps new ways to learn about each other may develop. It is possible that online companies could arise to profit from an "arms race" of new tactics to be both seen and hidden – in our age-old dance to maintain secrets, privacy, and relationships (Lepore, 2013).

Further exploration

1. Rank the following rights from most important to least important: in other words, which rights would you give up for other rights, if a decision was forced? Defend your ranking with reasons. The potential rights: confidentiality (shared personal information could not be disclosed by receiver), consent (explicit permission is required for your information to be used), forgotten (personal information can be removed from publicly accessible media), free speech/expression (you can say openly and publicly what you believe), free press (you can publish accurate accounts of events or issues), open records (you have a right to know information that could affect your life), privacy (there is a defensible line between what you know about yourself and what others can know about you). Are there other relevant rights you would include in your list?
2. How does the right to be forgotten affect future understandings of history?
3. What in your own life would you like to be forgotten and not part of a public record? What if everyone removed such information about themselves? How might that impact public safety or democracy?

References

Ball, J. (2014, May 14). Costeja González and a memorable fight for the "right to be forgotten." *The Guardian*. Retrieved from http://www.theguardian.com

Biersdorfer, J. D. (2015, May 20). Unlike you, your Facebook account can be immortal. *The New York Times*. Retrieved from https://www.nytimes.com

Bond, Jr., C. F., & DePaulo, B. M. (2006). Accuracy of deception judgments. *Personality and Social Psychology Review, 10*(3), 214–234.

Erlanger, S. (2013, October 3). Britain: Online surveillance challenged. *The New York Times*. Retrieved from https://www.nytimes.com

Essers, L. (2015, May 13). Google rejects 60 percent of right to be forgotten requests. *PC World*. Retrieved from http://www.pcworld.com

Garsd, J. (2015, March 3). Internet memes and "the right to be forgotten." National Public Radio. Retrieved from http://www.npr.org

Goffman, E. (1959). *The presentation of self in everyday life*. New York, NY: Anchor/Doubleday.

Goleman, D. (1996). *Vital lies, simple truths: The psychology of self-deception*. New York, NY: Simon & Schuster.

Golin, S., & Bregman, A. (Producers), & Gondry, M. (Director). (2004). *Eternal sunshine of the spotless mind* [Motion picture]. United States: Anonymous Content/This Is That.
Good, K. D. (2012). From scrapbook to Facebook: A history of personal media assemblage and archives. *New Media & Society, 0*(0), 1–17. Retrieved from http://infra.sarai.net/lib/files/original/4af69b75b39cd3ba68c83f0c7e30a726.pdf
Kauffmann, S. (2014, December 19). Google: Europe's favorite villain. *The New York Times*. Retrieved from https://www.nytimes.com
Lepore, J. (2013, June 24). The prism: Privacy in an age of publicity. *The New Yorker*, pp. 32–36
Mayer-Schonberger, V. (2009). *Delete: The virtue of forgetting in the digital age*. Princeton, NJ: Princeton University Press.
Opsahl, K., & Reitman, R. (2013, April 22). The disconcerting details: How Facebook teams up with data brokers to show you targeted ads [Web log post]. Electronic Frontier Foundation. Retrieved from http://www.eff.org
Parkinson, J. (2014, July 31). The perils of the Streisand effect. British Broadcasting Corporation. Retrieved from http://www.bbc.com
Rawlinson, K. (2015, May 13). Google in "right to be forgotten" talks with regulator. British Broadcasting Corporation. Retrieved from http://www.bbc.com
Rosen, J. (2012, February 13). The right to be forgotten. *Stanford Law Review Online, 64*, 88–92. Retrieved from http://www.stanfordlawreview.org/sites/default/files/online/topics/64-SLRO-88.pdf
Sayare, S. (2013, December 14). France broadens its surveillance power. *The New York Times*. Retrieved from https://www.nytimes.com
Scott, M. (2014, July 8). European companies see opportunity in the "right to be forgotten." *The New York Times*, p. B3. Retrieved from https://www.nytimes.com
Scott, M. (2015a, February 1). A question over the reach of Europe's "right to be forgotten" [Web log post]. *The New York Times*. Retrieved from https://bits.blogs.nytimes.com
Scott, M. (2015b, May 25). Who's the watchdog? In Europe, the answer is complicated [Web log post]. *The New York Times*. Retrieved from https://bits.blogs.nytimes.com
Shahani, A. (2014, July 14). In Europe, Google stumbles between free speech and privacy [Web log post]. National Public Radio. Retrieved from https://www.npr.com
Shahani, A. (2015, May 27). How will the next president protect our digital lives? [Web log post]. National Public Radio. Retrieved from http://www.npr.org
Smale, A. (2015, May 7). Germany limits cooperation with U.S. over data gathering. *The New York Times*. Retrieved from https://www.nytimes.com
Steinhauer, J., & Weisman, J. (2015, June 2). U.S. surveillance in place since 9/11 is sharply limited. *The New York Times*. Retrieved from https://www.nytimes.com
Tavris, C., & Aronson, E. (2007). *Mistakes were made (but not by me): Why we justify foolish beliefs, bad decisions, and hurtful acts*. Orlando, FL: Harcourt.
Tenner, E. (1997). *Why things bite back: Technology and the revenge of unintended consequences*. New York. NY: Vintage.
Todd, S., & Todd, J. (Producers), & Nolan, C. (Director). (2001). *Memento* [Motion picture]. United States: Summit Entertainment/Team Todd.
Toobin, J. (2014, September 29). The solace of oblivion. *The New Yorker*. Retrieved from http://www.newyorker.com

23
Virtual Currency

Money symbolizes value. In 2008, an anonymous computer programmer wrote open-source code that created the increasingly popular virtual currency, Bitcoin, free of government regulation, identification of buyers and sellers, and middlemen fees (Grinberg, 2011; Vigna & Casey, 2015). People can download the software to "mine" Bitcoins (Popper, 2015) and create their own virtual currencies (e.g., Ramos, 2014; Ward, 2014). Bitcoin is valuable for its ease of use, instantaneous international transfer of funds, and stability in countries with weak financial or property rights systems (Ember, 2015; Yu, 2014). And its own dollar value, perhaps surprisingly, temporarily skyrocketed (Phillips, 2013).

Despite these benefits, very little agreement – or even understanding – of Bitcoin exists. Is it a currency used to buy and sell stuff, or an investment traded for profit or loss, or property? From 2013 to 2015, governments debated but did not coordinate: Bitcoin is not a currency in China (Popper, 2013), is a currency in Japan, and is property in Singapore (BBC Staff, 2014g). In the US, different regulatory agencies compete over whether it is an investment (Popper, 2013), property (Peralta, 2014), or something else (Caesar, 2014). New York licenses Bitcoin exchanges (Merced, 2015).

With Bitcoin, we turn over our money and perhaps financial well-being to a computer program. How much trust do we place in computers? Is removal of human experts in financial transactions cause for concern or celebration? If something goes wrong – hacker "hold-ups," a software bug or computer glitch, or misdirected funds – who (or what) do we turn to for help? If the whole virtual currency network went awry, the lost value of the Great Depression may seem just a whimper in comparison. This case depicts real issues related to virtual currency in a fictional format. The event, names, and quotes are all fictitious.

The Value of Instant Anonymity

with Yonathan Bassal

Newscaster: This international panel on virtual currency follows government debates worldwide (e.g., Popper, 2013; Zak, 2014). Virtual currencies are not issued or backed by governments (Satran, 2013), but are increasingly accepted worldwide (Popper, 2015). We now go live to the discussion...

Secretary-General: Money pervades our lives. An innovative form – so-called virtual currency – is basically bits of computer code. What role will it have for the well-being of our people locally and globally? Thank you to our panelists addressing, from various perspectives, four ethical issues that virtual currencies evoke: anonymity, illegality, security, and volatility. With virtual currency, people in transactions do not have to disclose their personal identities, so it may be unclear who we are dealing with, which leads to a fertile environment for illegal transactions. Plus, software code can be hacked, so would our digital wallets be safe? And in the last few years, the value of the most well-known virtual currency, Bitcoin, has gone up and down wildly, causing fear and loss. Let's start with: How does virtual currency work?

Computer Programmer: Bitcoin was created by a pseudonymous coder. It is created through software, which anyone can download and run to become part of a worldwide network that "mines" Bitcoins (Nakamoto, 2008). The software solves equations that verify Bitcoin transactions, which once verified become part of a "block chain" ledger of every Bitcoin transaction ever. The computer that "completes" a block has "mined" a Bitcoin, which is stored in a digital wallet (Sawyer, 2013).

Ambassador: Thanks for the technical details. I would like to jump to ethics with a case: Mt Gox, a Tokyo-based Bitcoin exchange, which got huge media coverage in 2013–2014. A hacker took advantage of a bug in Bitcoin code and stole millions of dollars from Bitcoin's largest exchange, which contributed to Bitcoin's ongoing volatility (BBC Staff, 2014c). Who knows who's at fault? This security breach shut down the largest exchange and many people lost money (Villar, Knight, & Wolf, 2014). They fixed the technical glitch (BBC Staff, 2014e), but how can we trust our money going who knows where?

Law Enforcement Official: Mt Gox was not the only exchange hit by hackers and ended up folding (e.g., Kelion, 2014). I want to focus on virtual currencies' association with crimes. I think the focal case should be Silk Road, an exchange *designed* for illegal activity (BBC Staff, 2014d)! Everyone there used Bitcoin for anonymity, so it's hard to find the criminals (Kaplanov, 2012). When the FBI dismantled Silk Road, it confiscated and auctioned Bitcoins (BBC Staff, 2014b, 2014d, 2014k), but only a pittance of the amount circulating through illegal transactions (Satran, 2013). We need to keep illegal marketplaces off the grid.

Merchant: But most people don't go there, right? For regular commerce, you can't deny businesses are accepting Bitcoin (Ember, 2014a). There are many stores with "Bitcoin Accepted" signs in the window.

Ambassador: Politicians accept Bitcoin for their campaigns, too (Caesar, 2014; Lichtblau, 2015).

Merchant: My business accepts Bitcoin. More well-known companies are doing it (BBC Staff, 2014j; Ember, 2014a).

Law Enforcement Official: How do you know you aren't laundering money inadvertently (BBC Staff, 2014c)? You're hurting the hard-working, law-abiding citizens if we aren't protecting them from crime.

Computer Programmer: In some countries, Bitcoin is becoming *the* regular currency for consumers (Popper, 2015)!

Merchant: At the day-to-day level, Bitcoin makes things easier. People like it. Fewer fees (Ember, 2014a). I give customers discounts for using Bitcoin. There are even ATMs (Morisy, 2014; Pressman, 2013).

Consumer Advocate: If Bitcoin becomes common and integrated with apps (BBC Staff, 2014i), shopping (Lee, 2014), even pensions (Castronova & Fairfield, 2014), every consumer might be affected by volatility or security (Ember, 2014b). Could you imagine your grandparents losing their entire retirement as Bitcoin values drop or cyberthieves strike (Ward, 2014)? Governments need to make some decisions (BBC Staff, 2014h) and coordinate with each other (BBC Staff, 2014g; Popper, 2013). Everyone thinks Bitcoin is independent of governments, but its value responds to government actions (Popper & Gough, 2013) as much as investor speculation (BBC Staff, 2014a, 2014l) or technical issues (BBC Staff, 2014c).

Law Enforcement Official: Falling values actually keep crime at bay. When Bitcoin was hot, malware and ransomware were infecting unsuspecting people's computers (Perlroth & Wortham, 2014; Wakefield, 2014). If these currencies aren't worth much, cyberthieves are less interested (BBC Staff, 2015b).

Economic Adviser: I'm concerned virtual currencies might destroy our banking and financial services system. Not the Bitcoins themselves, but how they are created and tracked (Ember, 2015). If they become the norm, whole professions may suffer: accountants, lawyers, bankers could be replaced by block chain technology.

Consumer Advocate: If they go away, what happens if there is a problem or dispute?! Who would the consumer turn to (Ember, 2015)?

Law Enforcement Official: Or worse, imagine the system was hacked and all the digital money disappeared, what would be left? No government, no money backed by an institution. If Mt Gox was a regular American bank, deposits would've been insured (BBC Staff, 2014f; Lee, 2013).

Ambassador: Also, virtual currency makes it difficult to tax transactions. How are governments going to raise money for services?

Economic Adviser: Yes, let's not forget all the government employees, recipients of government funds like Social Security and Medicare/Medicaid, and college students with federal financial aid. If we can't trace Bitcoin transactions for taxes, then no more support a lot of people depend on.

Law Enforcement Official: After the Mt Gox hack, the exchange updated its version of the software to issue unique transaction identifiers (BBC Staff, 2014e). Maybe that would help. It doesn't identify the person, but does identify the transaction (Castronova & Fairfield, 2014). Maybe tax digital wallets and it doesn't matter who the person is.

Secretary-General: What I'm really thinking about now is feasibility. What is feasible for governments to do, now with all the people already involved with virtual currency? And what are individuals responsible for regarding their own behavior?

Economic Adviser: The younger generation is tech savvy, and the rest of us adapt. Investors and venture capitalists are interested, if wary (Briere, Oosterlinck, & Szafarz, 2013; Krugman, 2013), especially about security (Ember, 2015; Miller, 2014). Maybe a reputation system, like online auction and travel review sites have?

Computer Programmer: Bitcoin is controlled by supply – there are only so many that can be mined. It's not run by a person or a policy, which distinguishes it from cash. Regulation isn't needed in terms of following transactions. Those are trackable. And if one virtual currency fails, another one would gain momentum (Popper, 2014). We need to build in trust systems for *people*.

Economic Adviser: There are strong opinions on this innovative virtual currency. Actually, we don't really agree what Bitcoin is yet – a currency, property, investment (e.g., BBC Staff, 2014g)?

Researcher: That is the problem...we need to understand how it will work (BBC Staff, 2015a). And the experiment is occurring in the real world with real money, not studied in a lab far from financial markets. Knowledge might help us make better decisions. Time is of the essence, since virtual currency is already in use.

Secretary-General: Thank you for your expertise. Although it is scary, new, and disruptive, virtual currency has global implications and the chain effect that our decisions could have all around the world could have irreversible consequences. Virtual currency brings to light ethical gray areas in our financial system. Is it a fad or the future of money?

Further exploration

1. As a lawmaker, how would you address the pros and cons of virtual currency in your deliberations? A vote is called: would you support or reject virtual currency outright or only in particular circumstances? Defend your decision.
2. How is "value" constructed? How do the roles of virtual currency users, toward both legal and illegal ends, contribute to this value?
3. Imagine a world where virtual currency was the only form of payment. Describe that world.

References

BBC Staff. (2014a, January 6). Bitcoin crosses $1000 on Zynga move. British Broadcasting Corporation. Retrieved from https://www.bbc.com
BBC Staff. (2014b, January 17). Silk Road forfeits Bitcoins worth $28 million. British Broadcasting Corporation. Retrieved from https://www.bbc.com
BBC Staff. (2014c, February 10). Bitcoin value drops sharply after tech issues continue. British Broadcasting Corporation. Retrieved from https://www.bbc.com
BBC Staff. (2014d, February 14). Silk Road 2 loses $2.7m in Bitcoins alleged hack. British Broadcasting Corporation. Retrieved from https://www.bbc.com
BBC Staff. (2014e, February 18). Big Bitcoin exchange "fixes" trading glitch. British Broadcasting Corporation. Retrieved from https://www.bbc.com
BBC Staff. (2014f, March 5). Bitcoin bank closes down after $600000 hacker theft. British Broadcasting Corporation. Retrieved from https://www.bbc.com
BBC Staff. (2014g, March 6). Bitcoin not a currency says Japan government. British Broadcasting Corporation. Retrieved from https://www.bbc.com
BBC Staff. (2014h, March 11). New York regulator plans "regulated" Bitcoin exchanges. British Broadcasting Corporation. Retrieved from https://www.bbc.com
BBC Staff. (2014i, June 3). Apple warms to apps using virtual currencies. British Broadcasting Corporation. Retrieved from https://www.bbc.com
BBC Staff. (2014j, June 12). Expedia to accept Bitcoin payments for hotel bookings. British Broadcasting Corporation. Retrieved from https://www.bbc.com

BBC Staff. (2014k, June 13). US to auction seized Silk Road Bitcoins worth $18m. British Broadcasting Corporation. Retrieved from https://www.bbc.com
BBC Staff. (2014l, October 6). Bitcoin price falls to 11-month low. British Broadcasting Corporation. Retrieved from https://www.bbc.com
BBC Staff. (2015a, March 18). Budget pledges funds for digital cash research. British Broadcasting Corporation. Retrieved from http://www.bbc.com
BBC Staff. (2015b, April 24). Bitcoins "losing" value for cyber-thieves. British Broadcasting Corporation. Retrieved from https://www.bbc.com
Briere, M., Oosterlinck, K., & Szafarz, A. (2013). Virtual currency, tangible return: Portfolio diversification with Bitcoins. *Social Science Electronic Publishing, 1*. Retrieved from http://ssrn.com
Caesar, C. (2014, May 8). Bitcoins for political donations? Boston.com. Retrieved from http://www.boston.com
Castronova, E. & Fairfield, J. A. T. (2014, September 10). The digital wallet revolution. *The New York Times*. Retrieved from https://www.nytimes.com
Ember, S. (2014a, August 14). For merchants, Bitcoin shows more pop than potential [DealBook web log post]. *The New York Times*. Retrieved from http://dealbook.nytimes.com
Ember, S. (2014b, December 18). New York regulator outlines changes in Bitcoin rules [DealBook web log post]. *The New York Times*. Retrieved from http://dealbook.nytimes.com
Ember, S. (2015, March 2). Data security is becoming the sparkle in Bitcoin. *The New York Times*, p. B1. Retrieved from http://www.nytimes.com
Grinberg, R. (2011). Bitcoin: An innovative digital currency. *Hastings Science and Technology Law Journal, 4*(1), 159–208. Retrieved from http://papers.ssrn.com
Kaplanov, N. M. (2012). Nerdy money: Bitcoin, the private digital currency, and the case against its regulation. *Temple Law Review, 1*, 1–46. Retrieved from http://ssrn.com
Kelion, L. (2014, February 12). Bitcoin exchange halts withdrawals after cyber-attack. British Broadcasting Corporation. Retrieved from https://www.bbc.com
Krugman, P. (2013, April 15). The antisocial network of Bitcoins. *The New York Times*. Retrieved from http://www.nytimes.com
Lee, D. (2014, February 14). Self-updating Bitcoin price tag shown off in east London. British Broadcasting Corporation. Retrieved from https://www.bbc.com
Lee, T. B. (2013, May 15). The coming political battle over Bitcoin [web log]. *The Washington Post*. Retrieved from http://www.washingtonpost.com.
Lichtblau, E. (2015, April 9). In accepting Bitcoin, Rand Paul raises money and questions. *The New York Times*. Retrieved from https://www.nytimes.com
Merced, M. (2015, June 3). Bitcoin rules completed by New York regulator. *The New York Times*. Retrieved from https://www.nytimes.com
Miller, J. (2014, January 10). Bitcoin vault offering insurance is "world's first." British Broadcasting Corporation. Retrieved from https://www.bbc.com
Morisy, M. (2014, February 19). Bitcoin ATM pulls into Boston. Boston.com. Retrieved from http://www.boston.com
Nakamoto, S. (2008). Bitcoin: A peer-to-peer electronic cash system. *Bitcoin.org*. Retrieved Oct 21, 2013, from http://Bitcoin.org/Bitcoin.pdf
Peralta, E. (2014, March 25). IRS says it will treat Bitcoins as property, not currency [Two-Way web log post]. National Public Radio. Retrieved from https://www.npr.com

Perlroth, N., & Wortham, J. (2014, April 3). Tech start-ups are targets of ransom cyberattacks [Web log post]. *The New York Times*. Retrieved from https://bits.blogs.nytimes.com

Phillips, J. (2013, December 1). Virtual currency Bitcoin bursts through real $1,000 barrier. *NBC News*. Retrieved from http://www.nbcnews.com

Popper, N. (2013, December 5). In the murky world of Bitcoin, fraud is quicker than the law [DealBook web log post]. *The New York Times*. Retrieved from http://dealbook.nytimes.com

Popper, N. (2014, February 17). Regulators and hackers put Bitcoin to the test [DealBook web log post]. *The New York Times*. Retrieved from http://dealbook.nytimes.com

Popper, N. (2015, May 3). Quick change. *The New York Times Magazine*, pp. MM48–53, 78–80.

Popper, N., & Gough, N. (2013, December 18). Bitcoin, nationless currency, still feels governments' pinch [DealBook web log post]. *The New York Times*. Retrieved from http://dealbook.nytimes.com

Pressman, A. (2013, November 17). First ATM offering Bitcoins opens in Vancouver. Retrieved from http://finance.yahoo.com

Ramos, J. (2014, March 7). A Native American tribe hopes digital currency boosts its sovereignty. National Public Radio. Retrieved from http://www.npr.org

Satran, R. (2013, May 15). How did Bitcoin become a real currency? *US News and World Report*. Retrieved from http://money.usnews.com

Sawyer, M. (2013). The beginner's guide to Bitcoin: Everything you need to know. Monetarism – A UK Money and Personal Finance Blog. Retrieved October 21, 2013 from http://www.monetarism.co.uk

Vigna, P., & Casey, M. J. (2015). *The age of cryptocurrency: How Bitcoin and digital money are challenging the global economic order*. London, UK: St. Martin's Press.

Villar, R., Knight, S., & Wolf, B. (2014, February 25). Bitcoin exchange Mt Gox goes dark in blow to virtual currency. Reuters. Accessed through Yahoo News.

Wakefield, J. (2014, January 8). Yahoo malware enslaves PCs to Bitcoin mining. British Broadcasting Corporation. Retrieved from https://www.bbc.com

Ward, M. (2014, April 25). How to mint your own virtual money. British Broadcasting Corporation. Retrieved from https://www.bbc.com

Yu, A. (2014, January 15). How virtual currency could make it easier to move money [All Tech Considered web log post]. National Public Radio. Retrieved from https://www.npr.com

Zak, D. (2014, January 23). Bitcoin's next frontiers: ATMs and Capitol Hill. *The Washington Post*. Retrieved from http://www.washingtonpost.com

24
Emoticons

As we segue from face-to-face communication to digitally mediated communication, what do we do about the nonverbal dimensions of meaning? Raised eyebrows, a wink, or a crinkled nose are, as the cliché says, worth a thousand words. Emotional expression helps us navigate the social world, even online (Vandergriff, 2013). When the Internet was young, and users were stuck with only text, online forums became rowdy because of misunderstandings. A young tech guy cleverly realized that punctuation combinations could symbolize inner states and intentions – emoticons were born (Kennedy, 2012). In Japan, elaborate little cartoons, some with motion, developed – emojis, which are now integrated features of smartphones in the US (Wortham, 2011). Emoticons and emojis went viral as an easy way to convey joking or sarcasm (Garber, 2013).

These little guys have come a long way. They are one of the fastest growing languages (Doble, 2015; Isaac, 2015). Their use sometimes leads to confusion and trouble, and they need their own grammar (Bowman, 2015). Emojis may replace Internet slang like "LOL" and "OMG" (Isaac, 2015) as well as numeric passwords (Sanders, 2015). They have their own "government" – the Unicode Consortium – to make sure they display understandably across technology platforms (NPR Technology Staff, 2014). They are diversified by race (Kelion, 2014), gender, and body type (Engeln, 2015). Cultural differences have been found in emoji preferences (NPR Staff, 2015). Some consider these new options as an embrace of differences (Chow, 2015), but others read them as stereotypical and sometimes derogatory (Engeln, 2015; Peralta, 2015).

They grow sophisticated, conveying emotions much more complex than smiley and frowny faces (Sharrock, 2013). You might say they have their own museums: online venues where adoring fans publish stories written only using emojis (Narratives in emoji, n.d.). Emojis are even having their day in court – as central evidence in cases of cyberbullying and online criminal networks (Weiser, 2015). However, critics feel emoticons are overused or misused, and *symbolize the death of effective writing (Newman, 2011). Their simplicity betrays emotional immaturity (Haber, 2015), and they should be avoided in professional situations (Wortham, 2009) and early dating conversations (Haber, 2015).*

Eventually, emoticons and emojis might not need us humans. They can be used to further train digital devices to become more human-like (Hardy, 2015). Perhaps, one day, our devices might emit their own emojis, sharing with us or each other how they feel. How does that possibility make you feel? :-) or {:o or >:-< or perhaps a little ;-P

:-) or :-(?

with Jai Sung Lee

The widespread popularity and growing availability of laptops, tablets, and smartphones has turned us into a society of texters, update posters, and frienders. These gadgets are changing our use of language (Abu Sa'aleek, 2013). Our communication is less face to face, or even voice to voice, but rather fingertip to fingertip. "Text" has become a verb. We have started to change how we express and perceive meaning from these mediated interactions. How do we replace the nonverbal cues we were used to? Emoticons! Now, with a series of punctuation marks, we can succinctly convey our emotions to not only people we know but to strangers worldwide who follow our digital feeds.

These innocuous-looking symbols have a storied history. There is a general consensus that the inventor of the first emoticon must have knowingly utilized the symbol to convey the meaning that it has now (Dolak, 2012). General agreement converged on an electronic message that a professor wrote in 1982 to a computer science department bulletin board as the place where emoticons were born (Kennedy, 2012). In this message, the inventor not only presented the colon, dash and close-parenthesis as a smiley face, he also explained how to read it ("sideways"), and conveyed its purpose as "joke markers" (Long, 2008).

Since that first email, emoticons continue to evolve, becoming more colorful and complex. Now we have a wide variety, including cartoon characters in different poses (Houston, 2013).

Millennials – the generation of digital natives – take for granted that everyone can understand what emoticons mean. Whenever we see :-)

or :), we automatically see these symbols as smiling faces. However, not everyone sees eye-to-eye (or : 2 :) on emoticons, especially across generations. A fictional, intergenerational conversation demonstrates the ethical quandaries of emoticons...

It's near bedtime on a Friday night. Chris, a communications professor, walks into the living room and sees his teenage son, Paul, lounging on the sofa. Chris asks, "Why aren't you hanging out with your friends, Paul?" Paul replies, "I am." Chris sees that Paul is furiously typing away on his smartphone. Chris asks, "Is that how you talk with all your friends?" Paul merely nods.

Chris takes a deep breath and launches into a rant. He talks about how things have changed since his time, especially how texting, messaging, emailing, and social media posts have taken over. He bemoans the loss of *talk* – real talk! – and how many nuances and subtleties are lost with only typed correspondence. Chris quotes a book he read by Mehrabian (1972), about how nonverbal cues are more important than words. "Sixty percent – 60 PERCENT!" Chris raises his voice, "of understanding what someone says comes from faces and movement. How can you 'get the message' through texting?"

Paul's eyes remain glued to his smartphone, flickering with focused concentration every time his phone vibrates that a text message has arrived.

It doesn't matter, because Chris doesn't stop for an answer from Paul. "You know, it's actually worse than that because tone of voice carries a lot of the meaning, too. One of my colleagues, Thompson (2011), estimates that words account for very little." With an air of triumph, Chris exclaims, "So for kids in your generation, you guys are missing out on up to 93 percent of what's important in a conversation! Talk about inefficiency...," he trails off, shaking his head.

At this last statement, Paul briefly looks up at Chris, but then silently resumes typing. Seconds later, Chris hears his cellphone ring in the other room. Curious who would be calling him at this late hour, Chris looks for his phone. He sees that Paul sent him a text message. Intrigued, Chris opens the message to see a series of symbols: >:-(

Chuckling, Chris walks back into the living room to see Paul look at him with a smug expression: "That was pretty efficient, huh, dad? I didn't even have to say anything, yet you knew what I was feeling. As a matter of fact, I didn't even have to send you *any* words to portray my feelings. Emoticons are the way of the future, dad. They are so simple, yet effective. Talk about efficiency!"

Never one to back down from a challenge, especially with his son, Chris concedes, "I'll give it to you. Emoticons are pretty efficient. But, just because an invention is more efficient does not make it better, or that the older version should be replaced by the new. Take violins, for example. Modern violin producers can manufacture multiple violins a day. However, the world's most sought after and most expensive violins are those handcrafted way back in the 1600s" (Saunders, 2011).

Chris paused, then added for emphasis, "Emoticons are a one-trick pony."

Paul tilts his head to the side, with one of his eyebrows up.

Chris continues, "Yup, emoticons can only be used in an informal setting. In the real world – the grown-up world – almost no one uses them." He smirks, "You will grow out of emoticons, just like you will stop wearing those skinny jeans."

Chris is startled as Paul types furiously into his smartphone, and Chris's phone vibrates. The screen lights up: "Too bad you couldn't convey that sassy sarcasm through your grown-up writing. I can do it in three keystrokes: ;-P "

Paul turns toward Chris, "Dad, I'm not sure why you are so against emoticons. Most people are against change only because it will change the way things were, and they can't come to a realization of a dynamic world. There are so many positives to emoticons, and they have so much potential to completely change communication. You're just not used to them...and afraid that emoticons will replace the formal writing you are so accustomed to!"

Chris tries to imagine a world where emoticons are the norm. He laughs aloud as he thinks about an email his boss sent him earlier about an upcoming meeting. He pulls up the email on his laptop: "Dear Chris, Please prepare the presentation for our board of trustees next week. Be sure to avoid the topic of the recent loss of our colleagues, as that could spell disaster for us with next year's budgeting."

He looks up to Paul's perplexed expression and says, "Can you imagine what would happen if my boss had used emoticons in this email? I don't think I could have read this email without laughing as I am now."

Paul read the email and grinned sheepishly. "You're right, dad. It would seem a little funny if he used emoticons in that email. But think about it. It seems funny now because it *isn't* the norm. But imagine if emoticons become common. I mean, emoticons...they set the tone...they save time...a few symbols and you would know exactly what he is looking for. And as you always say, time is money, so saving time is saving money. Let's face it, :-) is so much simpler and more heartfelt than 'I am happy' – and cuter, too."

Chris leans in close to look Paul square in the eyes, "What if emoticons *did* become the norm? What kind of world would we be living in? Sure, people could express their emotions on paper. Let's say that emoticons completely take over writing. Maybe some people start to think: 'Hey, why waste time learning how to write my thoughts with words? I can just use emoticons instead.' What's going to happen to literacy if no one can read or write words? What's going to happen to literature, to public documents and other records? Who will make sure the past doesn't completely disappear from memory? Hmm?"

Chris pauses to take a breath. "If emoticons became the norm, and someone didn't add an emoticon to a message or an email, is he being rude? What would that suggest or imply? Let's take this whole line of thinking one step further. What if people become *solely* reliant on emoticons for communication? Will communication become more efficient – or perhaps less? Emoticons cannot offer many subtleties."

Paul took a minute to let his dad's words sink in. "That's a great point, dad. But I think the beauty of emoticons is that they don't need to replace words but just enhance them. Think about the benefits. Like facial expressions, emoticons are universal, so you can make friends worldwide. You could use them when you travel to conferences" (Russell, 1994).

"Remember that time when we went to Spain?" Paul got excited at the memory. "No one spoke Spanish. But we could show we were friendly by smiling, and ask directions by pointing. Remember when you mistakenly starting rooting for the rival team at the soccer stadium? We didn't need to understand their words to realize they were none too happy with us. The scowls said it all, and we got out of there fast! Think, dad. It's like when Edison invented the light bulb and forever changed the lives of millions of people all around the world, even to this day! Emoticons light up feelings in digital media."

Chris, astounded at the loquaciousness of his usually quiet son, pondered, "Perhaps you're right, Paul. Emoticons do have potential for changing people's lives for the better. But are you sure the benefits outweigh the risks? People can't hide or fake real facial expressions (Russell, 1994). You can usually tell when someone is genuinely smiling. With emoticons, it's different. They can be manipulated ... then they can manipulate you. Emoticons can lie. What will happen to trust?"

Chris paused a moment to maintain his composure. "I'm glad you mentioned Spain – that was a great trip. But, as you recall, Spain differed

a lot from Russia, remember? Cultures vary. What about cultures that do not condone expressing emotion or see expressing emotions as a sign of weakness? What would happen if you emoted in a text there?"
Paul tired of this conversation. Plus, it was bedtime. He got up to leave. "Hey, dad?" "Yes?" "Thanks for the talk. I was all for emoticons, but I enjoyed this back and forth, hearing you get excited about the whole thing. Good night, dad."
As Paul turned off his bedroom light, his phone vibrated: "Great talking to you too, son. Love, Dad :-) "

Further exploration

1. Who – Chris or Paul – do you agree with more? Why? What are additional arguments that could support your side?
2. Create a list of situations in which emoticons are acceptable and not acceptable. What characteristics of a situation make emoticons acceptable?

References

Abu Sa'aleek, A. O. (2013). Linguistic dimensions of initialisms used in electronic communication. *Studies in Literature and Language, 6*(3), 7–13. Retrieved from http://scholar.google.com

Bowman, E. (2015, May 4). As emoji spread beyond texts, many remain [confounded face] [interrobang]. National Public Radio. Retrieved from http://www.npr.org

Chow, K. (2015, February 25). African emoji CEO: Apple "missed the whole point" with its diverse emojis. National Public Radio. Retrieved from http://www.npr.org

Doble, A. (2015, May 19). UK's fastest growing language is...emoji. British Broadcasting Corporation. Retrieved from https://www.bbc.com

Dolak, K. (2012, September 19). Emoticons turn 30: A brief history [ABC News web log]. Retrieved from http://abcnews.go.com

Engeln, R. (2015, March 13). The problem with "fat talk." *The New York Times*. Retrieved from https://www.nytimes.com

Garber, M. (2013, September). The way we lie now. *The Atlantic*, pp.15–16.

Haber, M. (2015, April 3). Should grown men use emoji? *The New York Times*, p. D27. Retrieved from http://www.nytimes.com

Hardy, Q. (2015, March 26). Facebook's Yann LeCun discusses digital companions and artificial intelligence. *The New York Times* [Bits web log]. Retrieved from http://bits.blogs.nytimes.com

Houston, K. (2013, September 27). Smile! A history of emoticons. *Wall Street Journal*. Retrieved from http://online.wsj.com

Isaac, M. (2015, May 1). The rise of emoji on Instagram is causing language repercussions [Bits web log]. *The New York Times*. Retrieved from https://www.nytimes.com

Kelion, L. (2014, March 26). Apple seeks greater emoji racial diversity. British Broadcasting Corporation. Retrieved from https://www.bbc.com

Kennedy, P. (2012, November 23). Who made that emoticon? *The New York Times Magazine*, p. MM20. Retrieved from https://www.nytimes.com

Long, T. (2008, September 19). Can't you take a joke? :-) [Web log]. *Wired*. Retrieved from http://www.wired.com

Mehrabian, A. (1972). *Nonverbal communication*. New Brunswick, NJ: Aldine Transaction.

Narratives in emoji. (n.d.). Retrieved from http://narrativesinemoji.tumblr.com

Newman, J. (2011, October 21). If you're happy and you know it, must I know, too? *The New York Times*. Retrieved from https://www.nytimes.com

NPR Staff. (2015, April 27). Canadians love poop, Americans love pizza: How emojis fare worldwide. National Public Radio. Retrieved from http://www.npr.org

NPR Technology Staff. (2014, June 30). Why 140 characters, when one will do? Tracing the emoji evolution. National Public Radio. Retrieved from http://www.npr.org

Peralta, E. (2015, February 23). In beta release, Apple introduces new, racially diverse emojis. National Public Radio. Retrieved from http://www.npr.org

Russell, J. A. (1994). Is there a universal recognition of emotion from facial expressions? A review of the cross-cultural studies. *Psychological Bulletin, 115*, 102–114.

Sanders, S. (2015, June 15). Emoji passwords could be coming your way. Is that a good thing? National Public Radio. Retrieved from http://www.npr.org

Saunders, E. (2011, June 21). What makes the Stradivarius violin so special? British Broadcasting Corporation. Retrieved from http://www.bbc.com

Sharrock, J. (2013, February 2). How Facebook, a Pixar artist, and Charles Darwin are reinventing the emoticon. *Buzzfeed*. http://www.buzzfeed.com

Thompson, J. (2011, September 30). Is nonverbal communications a numbers game? [Web log post]. Retrieved from http://www.psychologytoday.com

Vandergriff, I. (2013). Emotive communication online: A contextual analysis of computer-mediated communication (CMC) cues. *Journal of Pragmatics, 51*, 1–12. Retrieved from http://www.sciencedirect.com

Weiser, B. (2015, January 28). At trial, lawyers fight to include evidence they call vital: Emoji. *The New York Times*, p. A22. Retrieved from http://www.nytimes.com

Wortham, J. (2009, November 2). Tweens on Facebook, and emoticon overload [Gadgetwise web log]. *The New York Times*. Retrieved from https://www.nytimes.com

Wortham, J. (2011, December 6). Whimsical texting icons get a shot at success. *The New York Times*. Retrieved from https://www.nytimes.com

25
Digitally Mediated Communication

Mediation involves an intermediary that conveys information between two entities. Mediation is not new: there remain human mediators like peace negotiators, realtors, and lawyers who broker deals, and analog mediators like writing, printing, and landline phones. Now, our lives feel dominated by digital gadgets that mediate our social interactions – email, texting, social media, and apps (Planet of the phones, 2015).

Mediated communication can vary on three dimensions: anonymity, transparency, and synchrony. In some interactions, we know who we're talking to. But in others, we don't – and the other "person" could be software not human. In some interactions, the use of information exchanged is understood by both parties. But in others, it is unclear how that information might be used or repurposed. In some interactions, we receive responses in real time, which provides better feedback to us. But in others, responses may be delayed or may come from database information that was created years ago.

For some, mediated communication is the preferred way to communicate. For example, it can be advantageous for people who are shy, introverted, on the autism spectrum, or not speakers of the local language because mediation provides time, distance, and control of messaging. But even within long-term relationships, its use has risen: mothers who text children that dinner is ready and family meals eaten silently in the bluish glow of all the screens at the table (Tierney, 2008). Are we switching from gadgets distracting us from personal relationships to people distracting us from our gadgets?

With mediated communication, we feel in control over how others see us and perhaps less vulnerable as we hide behind our screens. Rather than the "messiness" of in-person conversation (Garber, 2014), mediated messages can be edited and sent on our command. Voyeuristically, we consume the posted diaries and videos of each other's lives. Are we blurring the ability to distinguish between our curated (fictional) online characterization of ourselves and the development of our real character (see Dalbudak, Evren, Aldemir, Coskun, et al., 2013)?

Usually, mediators are considered a neutral party – they transfer information, but don't create it. But what if the mediator has a vested interest in the outcome? Recent media stories suggest that the algorithms that run search engines, dating sites, and social media can manipulate information as it is transmitted (Goel, 2014; Wood, 2014), and they can have biases (NPR Staff, 2015b). Might these opaque algorithms change the way we relate to each other – not only online but perhaps even in person?

I Feel So Close to ... Who Are You?
with Michaela Hession

As electronic gadgets proliferate, we increasingly communicate with each other mediated by technology (Feiler, 2015). This growing mediation presents several ethical benefits as it bolsters our ability to connect in ways previously not possible. Yet, ethical challenges arise as it disrupts the feedback loops that reinforce good two-way communication and might change our understanding of communication to be only one-way messages. That is, while our gadgets might free us from constraints of geographical distance or editorial gatekeepers (NPR Staff, 2015a), they also might impede our communication skills (Bilton, 2014a). As a result, what if eventually we *only* communicate through media and never see each other in person? And what happens if hi-tech mediators take on self-interested identities of their own?

Empowering the vulnerable

Mediated communication quickly can distribute messages and coordinate actions invaluable to political and social reform (e.g., Barry, 2009). Cell phones and the Internet have enabled economic growth in locations too isolated to foster exchanges (Corbett, 2008; Shaffer, 2013). Apps allow someone in a dangerous situation – such as journalists, political activists, and abuse victims – to discreetly message for help (e.g., Chozick, 2012; McDonough, 2013). Clients can text their therapists (Singh, 2014), and the sick can seek healthcare without leaving home (Shahani, 2015).

Bridging distances

Numerous websites unite people with similar interests, ailments, problems, or goals. Crowd-sourcing sites provide us quickly with opinions, start-up funds, home remedies, or advice. Online courses allow students to learn just about anything from anywhere (Selingo, 2014). We can text nurses, describe or photograph our symptoms, and receive our prescriptions by cell phone (Shahani, 2015), freeing office visits for more serious situations.

When our families are far-flung, friends deploy in the military, or coworkers move to an international office, we can still inexpensively talk to them every day through the Internet. Not only are we able to hear Grandma's voice even though she lives 3,000 miles away, but we also can see her smile via video chat.

Maximizing work opportunities

Mediated communication expands job seeking opportunities (Workers of the world, log in, 2014). Professional networking sites make our resumés widely accessible to potential employers. This mediated job market forces us to categorize our work experiences and skills into keywords optimized for search engines rather than promoting our unique personal strengths.

Once we get a job, many of us may end up looking at screens more than interacting with people (Carr, 2014). Gadgets may mediate how employees collaborate (Manjoo, 2015) or even become our bosses (Wall, 2014). Tech-mediation transforms not only manufacturing jobs, but also professional positions, such as medicine and teaching (Tufekci, 2015).

Several companies have started tele-doctoring services (Shahani, 2015). However, the limited information available through mediated communication may decrease doctors' sensitivity to important cues for diagnosis and treatment (Wachter, 2015). Massive open online courses (MOOCs) make education inexpensive and widely available, but their reach and success have been disappointing (Breslow, Pritchard, DeBoer, Stump, et al., 2015; Selingo, 2014), they underappreciate the contributions of social interactions (Pinker, 2015), and they may destroy traditional universities (Cusumano, 2013).

Although considered convenient and efficient, mediated work's loss of contextual and social cues found in personal interactions may increase misunderstandings and reduce human productivity. The tech boom isn't booming as much as expected (Krugman, 2015). Advocates of mediated work say computer productivity will free us to pursue more complex, strategic, and creative work (Davenport & Kirby, 2015) or unchain us from offices (Shahani, 2015). Critics worry that the employees who have semi-automated jobs will be laid off as automation increases (Tufekci, 2015).

Expanding social pressures

Mediated communication may increase our connectedness, but is more necessarily better communication? Mediated communication makes it so easy to keep up with each other that it can feel like a social slight if a response is not speedy (Rosman, 2014). We feel pressure to check updates 24/7 (Soper, 2014), and we fret over our posts to avoid rejection, ridicule, or shaming (Reiner, 2013). Cyberbullying has proliferated (Duggan, 2014). Even well-intentioned people have joined anonymous, cruel online mobs (Bittner, 2015; Marche, 2015), eventually realizing that behavior turned them into someone they didn't like (Miller & Spiegel, 2015).

Many children born in the 21st century have never lived without tech-mediated communication, and not having their devices could cause them stress and anxiety (Dalbudak et al., 2013). Adults check their phones first thing each day, when using the toilet, and during sex (Planet of the phones, 2015). With devices as constant companions, we don't know what to do with ourselves when alone (Wilson, Reinhard, Westgate, Gilbert, et al., 2014).

Lying through our screens

We tend to disclose more information through mediated communication than we do in person, partly because we can't provide nonverbal clues and partly because we feel shielded by anonymity (Schouten, Valkenburg, & Peter, 2009). With all that disclosure aggregated into databases, it's possible that social media sites know us better than our families do (Murphy, 2014). Except what we tell each other may not be true. To garner ever more attention (Marche, 2015), keep "likes" and "retweets" coming (Mullainathan, 2014), and be "swiped right" (Toma & Hancock, 2012), we distort our posts. Deceit is prevalent online across users (Garber, 2013) and even by site algorithms (BBC Technology Staff, 2015).

Yet, mediated communication also increases the chance of getting caught lying – anyone can search for disconfirmation (Garber, 2013). Some social media are developing software to test the truthfulness of posts (BBC Staff, 2014). In person, we have feedback from others' faces and behaviors to help us decipher meaning and intent. But if our interactions become increasingly mediated, providing us fewer opportunities to learn the meanings of facial expressions or gestures, will we lose this face-to-face "lie detector" advantage? What if growing distrust from online deception affects trust in face-to-face interactions?

Gaining reach but losing touch?

Selfies, emojis, and thumbs-up only go so far in conveying feelings (Marche, 2015). As our social manners become more tuned to online venues, might we lose our ability to regulate emotions and empathize (Feiler, 2015)? Some worry that our mediated chatting *at* each other may destroy conversations *with* each other because we lose the patience and interest for real-time talk (Garber, 2014). Tailored for brevity, popularity, and searchability, online posts are more convenient than the back-and-forth of conversation.

Online sharing doesn't make us happier (Korss, Verduyn, Demiralp, Park, et al., 2013) and we crave hearing others' voices (Wortham, 2014). But we shy away from real relationships because they are more time-consuming, demanding, and uncertain (Wayne, 2014), and they make us feel more vulnerable (Reiner, 2013) than mediated interactions.

Becoming obsolete?

As artificial intelligence and robotics advance, it may become difficult to tell when we're communicating *with* a gadget – rather than *through* a gadget to another person. That is, the other person's role has been taken over by the gadget. For example, companies are testing first-aid responders (Sydell, 2015), surgeons (Wachter, 2015), hotel concierges (Keane, 2015), retail clerks (Hu, 2015), telemarketers, and border guards (Tufekci, 2015) that are robots or automated. These scenarios go beyond mediation because the gadget or software no longer is only a go-between conveying information, it is also the producer of the information. It's like automated phone support or self-checkout except we can't press "0" or "help" to get to a real person.

Tech-mediation could replace not only work roles but also other roles. We can create, design, or train our own idealized friends, playmates, and love interests online – no need for the real people to exist (Bilton, 2014b; Garsd, 2015; Hardy, 2015). These fake relationships can make us look popular, give us just the advice we want, and never need anything in return. When we tire of them, they're deleted.

Or perhaps fake friends won't be needed. The gadget itself might provide a sense of friendship (Morris & Aguilera, 2013) or self-therapy (Yuen, Goetter, Herbert, & Forman, 2012). Perhaps the extreme of mediated communication – such as virtual reality simulations – might allow us to feel supported and to work through our problems without friends or confidants (Hogenboom, 2014).

The most extreme scenario of mediated communication is: not only the other "person" becomes part of the technology, but so do we. We, as embodied individuals, become unnecessary. Virtual reality turns the mediator – the software – into the situation, and turns what is mediated – us and other people – into storylines. Our selves are avatars. Our lives are simulations. Currently, virtual reality systems cause nausea (Nelson, 2014). But once this physical discomfort issue is overcome, and once we believe we have true "presence" within the virtual world (Heffernan, 2014), will we "upload" our consciousness and live there permanently as "flowing information," as science fiction has surmised (Silver & the Wachowskis, 1999)?

Further exploration

1. In the future, if the *only* way we were allowed to communicate was through gadgets, what non-gadget ways of communication do you think you would miss most? Why?
2. Think of one type of relationship that exists now (such as parent-child, teacher-student, spouses, friends, government leader-citizen, journalist-informant, etc.). Write a short story that depicts what that relationship might be like if the two people could only interact through digital media.

References

Barry, E. (2009, April 7). Protests in Moldova explode, with help of Twitter. *The New York Times*. Retrieved from http://www.nytimes.com

BBC Staff. (2014, February 19). Lie detector on the way to test social media rumours. British Broadcasting Corporation. Retrieved from https://www.bbc.com

BBC Technology Staff. (2015, March 26). Tinder prank "tricked men into flirting with each other." British Broadcasting Corporation. Retrieved from http://www.bbc.com

Bilton, N. (2014a, October 22). Meet Facebook's Mr. Nice. *The New York Times*, p. E5. Retrieved from https://www.nytimes.com

Bilton, N. (2014b, November 19). Social media bots offer phony friends and real profit. *The New York Times*, p. E5. Retrieved from https://www.nytimes.com

Bittner, J. (2015, March 25). Twitter's broken windows. *The New York Times*. Retrieved from http://www.nytimes.com

Breslow, L., Pritchard, D. E., DeBoer, J., Stump, G. S., Ho, A. D., & Seaton, D. (2015, Summer). Studying learning in the worldwide classroom: Research into edX's first MOOC. *Research and Practice in Assessment, 8*, 13–25.

Carr, N. (2014). *The glass cage*. New York, NY: Norton.

Chozick, A. (2012, November 30). For Syria's rebel movement, Skype is a useful and increasingly dangerous tool. *The New York Times*. Retrieved from http://www.nytimes.com

Corbett, S. (2008, April 13). Can the cellphone help end global poverty? *The New York Times*. Retrieved from http://www.nytimes.com

Cusumano, M. A. (2013, April). Are the costs of "free" too high in online education? *Communications of the ACM, 56*(4), 1–4.

Dalbudak, E., Evren, C., Aldemir, S., Coskun, K. S., Ugurlu, H., & Yildirim, F. G. (2013). Relationship of internet addiction severity with depression, anxiety, alexithymia, temperament and character in university students. *Cyberpsychology, Behavior, and Social Networking, 16*(4), 272–278.

Davenport, T. H., & Kirby, J. (2015, June). Beyond automation. *Harvard Business Review*, pp. 58–65.

Duggan, M. (2014, October 22). Online harassment. Pew Research Center. Retrieved from http://www.pewinternet.org

Feiler, B. (2015, April 19). The eye-to-eye challenge. *The New York Times*, p. ST2. Retrieved from http://www.nytimes.com

Garber, M. (2013, September). The way we lie now. *The Atlantic*, pp.15–16.

Garber, M. (2014, January/February). Saving the lost art of conversation. *The Atlantic*. Retrieved from www.theatlantic.com

Garsd, J. (2015, February 13). This Valentine's Day, I'm loving the boyfriend I built for myself [All Tech Considered web log post]. National Public Radio. Retrieved from http://www.npr.org/blogs

Goel, V. (2014, July 2). After uproar, European regulators question Facebook on psychological testing [Web log post]. *The New York Times*. Retrieved from https://bits.blogs.nytimes.com

Hardy, Q. (2015, March 26). Facebook's Yann LeCun discusses digital companions and artificial intelligence [Bits web log post]. *The New York Times*. Retrieved from http://www.nytimes.com

Heffernan, V. (2014, November 16). Seeing is believing. *The New York Times*, pp. MM52–60. Retrieved from http://www.nytimes.com

Hogenboom, M. (2014, August 22). Study creates "time travel" illusion. British Broadcasting Corporation. Retrieved from http://www.bbc.com

Hu, E. (2015, May 14). She's almost real: The new humanoid on customer service duty in Tokyo. National Public Radio. Retrieved from http://www.npr.org

Keane, J. (2015, June 2). Robot check-in: The hotel concierge goes hi-tech. British Broadcasting Corporation. Retrieved from: http://www.bbc.com

Korss, E., Verduyn, P., Demiralp, E., Park, J., Lee, D. S., Lin, N., Shablack, H., Jonides, J., & Ybarra, O. (2013, August 14). Facebook use predicts declines in subjective well-being in young adults. *PLoS One, 8*(8), e69841. doi: 10.1371/journal.pone.0069841

Krugman, P. (2015, May 25). The big meh. *The New York Times*. Retrieved from https://www.nytimes.com

Manjoo, F. (2015, March 12). An office messaging app that may finally sink email. *The New York Times*, p. B1.

Marche, S. (2015, February 15). The epidemic of facelessness. *The New York Times*, p. SR1.

McDonough, K. (2013, November 21). This incredibly smart domestic violence app could save women's lives. *Salon*. Retrieved from http://www.salon.com

Miller, L., & Spiegel, A. (2015, February 13). Can a computer change the essence of who you are? [Shots web log post]. National Public Radio. Retrieved from http://www.npr.org/blogs

Morris, M. E., & Aguilera, A. (2013, July/August). Smarter phones, smarter practice. *Monitor on Psychology*, pp. 58–63.

Mullainathan, S. (2014, July 1). Why computers won't be replacing you just yet. *The New York Times*. Retrieved from https://www.nytimes.com

Murphy, K. (2014, October 4). We want privacy, but can't stop sharing. *The New York Times*, p. SR4. Retrieved from https://www.nytimes.com

Nelson, N. (2014, August 10). Virtual reality's next hurdle: Overcoming "sim sickness" [All Tech Considered web log post]. National Public Radio. Retrieved from https://www.npr.com

NPR Staff. (2015a, May 20). Debate: Is smart technology making us dumb? National Public Radio. Retrieved from: http://www.npr.org

NPR Staff. (2015b, June 7). What makes algorithms go awry? National Public Radio. Retrieved from: http://www.npr.org

Pinker, S. (2015, January 30). Can students have too much tech? *The New York Times*, p. A27. Retrieved from http://www.nytimes.com

Planet of the phones. (2015, February 28). *The Economist*. Retrieved from http://www.economist.com

Reiner, A. (2013, November 1). Looking for intimacy in the age of Facebook. *The New York Times*, p. ED38. Retrieved from https://www.nytimes.com

Rosman, K. (2014, November 9). Hiding in plain e-sight. *The New York Times*, p. S2.

Schouten, A. P., Valkenburg, P. M., & Peter. J. (2009). An experimental test of processes underlying self-disclosure in computer-mediated communication. *Journal of Psychological Research on Cyberspace*, 3(2), 1–15.

Selingo, J. J. (2014, October 29). Demystifying the MOOC. *The New York Times*. Retrieved from https://www.nytimes.com

Shaffer, R. (2013, December 4). Mobile payments gain traction among India's poor. *The New York Times*. Retrieved from http://www.nytimes.com

Shahani, A. (2015, April 30). The doctor will video chat with you now: Insurer covers virtual visits. National Public Radio. Retrieved from http://www.npr.org

Silver, J. (Producer), & The Wachowskis (Directors). (1999). *The matrix* [Motion picture]. United States: Village Roadshow Pictures and Silver Pictures.

Singh, M. (2014, June 30). Online psychotherapy gains fans and raises privacy concerns [Health web log post]. National Public Radio. Retrieved from http://www.npr.org/blogs

Soper, J. (2014, April 1). Study: Americans spend 162 minutes on their mobile device per day, mostly with apps. Retrieved from http://www.geekwire.com

Sydell, L. (2015, March 18). SXSW debuts robot petting zoo for a personal peek into the future [All Tech Considered web log post]. National Public Radio. Retrieved from http://www.npr.org/blogs

Tierney, J. (2008, January 23). Gadget addiction. *The New York Times*. Retrieved from http://www.nytimes.com

Toma, C. L., & Hancock, J. T. (2012). What lies beneath: The linguistic traces of deception in online dating profiles. *Journal of Communication, 62*, 78–97.

Tufekci, Z. (2015, April 19). The machines are coming. *The New York Times*, p. SR4. Retrieved from http://www.nytimes.com

Wachter, R. (2015). *The digital doctor*. New York, NY: McGraw-Hill.

Wall, M. (2014, October 9). Could a big data-crunching machine be your boss one day? British Broadcasting Corporation. Retrieved from https://www.bbc.co.uk

Wayne, T. (2014, November 9). Swiping them off their feet. *The New York Times*, p. S14.

Wilson, T. D., Reinhard, D. A., Westgate, E. C., Gilbert, D. T., Ellerbeck, N., Hahn, C., Brown, C. L., Shaked, A. (2014, July 4). Just think: The challenges of the disengaged mind. *Science, 345*(6192), 75–77. doi: 10.1126/science.1250830

Wood, M. (2014, July 18). OKCupid plays with love in user experiments. *The New York Times*. Retrieved from https://www.nytimes.com

Workers of the world, log in. (2014, August 16). *The Economist*. Retrieved from https://www.theeconomist.com

Wortham, J. (2014, September 21). Pass the word: The phone call is back. *The New York Times*, p. SR5.

Yuen, E. K., Goetter, E. M., Herbert, J. D., & Forman, E. M. (2012). Challenges and opportunities in internet-mediated tele-mental health. *Professional Psychology: Research and Practice, 43*(1), 1–8.

Index

abortion, 98–9
accountability, 9
acetaminophen, 121
addiction, 111–15
alcohol, 121
alienation, 159–60
anticipation, 19–21, 27
antidepressants, 123
anxiety, 17, 29
artificial intelligence, 186, 218
arts, 18–19
authenticity, 157–62
autonomy, 93
average, 174–9

Barmack, Joseph Ephraim, 146
Bhutan, 139
Big Data, 182–6
Bitcoin, 200–4
blame, 9
body-by-design industry, 97, 99–100
boredom avoidance, 145–51

caffeine, 121
carbon dioxide removal, 39
cars, driverless, 53–9, 64
chopsticks, 21
classrooms, 149
climate change, 37–42
clocks, 21
collectivist cultures, 158, 160
collywobbles, 28–9
common good, 18–19, 27, 130, 158, 159–60, 161–2
communities, 160, 161
complacency, 140
conformity, 174
constructive creativity, 27
contentment, 140
coping skills, 140–1
correlations, 183

creativity
  bias against, 16–17
  boredom and, 147, 150–1
  common good and, 18–19
  constructive, 27
  cultural development and, 4–8
  descriptions of, 3–4
  ethics and, 9–11
  expense of, 8
  levels of, 4–5
  meaning-making and, 6–7
  moral dimension of, 10–11
  practicing by example, 29–30
  production of, 27–8
  relationships and, 8
  as systemic, 5–6
  time and, 8, 19–20
  worries about, 15–17
creators, 16, 162
cultural development, 4–8, 25
cultural imperialism, 139–40
culture
  definition of, 6
  self and, 158
customs, 6

Darwin, Charles, 93–4
data, 182–6
data protection laws, 193–8
daylight savings time, 21
developmental theory, 5
digitally mediated communication, 207–12, 214–19
disruptive innovation, 8, 27
distraction addiction, 147
diversity, 178–9
dopamine, 122, 123
driverless cars, 53–9, 64
drones, 69–74
  appeal of, 74
  military use of, 70
  personal safety and, 70–1

drones – *continued*
 privacy and, 71–2
 prosocial perspectives on, 72–3
 security and, 72
early adopters, 7, 16
e-cigarettes, 111–15
electronic cigarettes, 111–15
electronic devices, 147–8, 214–19
emojis, 207–8
emoticons, 207–12
emotions
 anticipation of, 20
 disconnection from, 148
 genuine, 122
 manipulation of, 121–7
 mediated communication and, 218
endosymbiosis theory, 90–1
Escher, M.C., 11
ethical anticipation, 19–21, 27
ethical dilemmas, 10
ethical possibilities, 28
ethics, 18–19
 anticipation and, 20
 creativity and, 9–11
 definition of, 9
 eudaimonic happiness, 141–2

Fairey, Shepard, 18–19
Feldman, David, 19
feminism, 167
fetuses, use of, in stem cell research, 98–9
fitting in, 175
food deserts, 108
forecasting, 19–20
foresight, 19–20
forks, 21–2
fortified junk food, 104–8
The Fountainhead (Rand), 3
Frankenstein (Shelley), 3
future, 19–21

gender, 165
gender fluidity, 165–70
gender identity disorder, 165
gene testing, 81–7
genius, 4
geoengineering, 37–42
 disincentives for, 41
 ethical implications, 39–42
 inequalities and, 41–2
 types of, 39
 weak regulatory momentum and, 40–1
germline engineering, 97, 100–1
The Giver (Lowry), 3–4
global warming, 37–42
government, 150, 195
Greek mythology, 3, 137–8
Gruber, H.E., 5
Guerilla Girls, 19
gut bacteria, 89–94

happiness, 137–42
 comparisons and, 139
 contentment and, 140
 coping skills and, 140–1
 eudaimonic, 141–2
 hedonic, 141–2
 money and, 138
 promotion of, 139–40
hedonic happiness, 141–2
Hierarchy of Needs, 158
history, 197
hope, 17
human control, 90–1
Human Genome Project, 82, 89
Human Microbiome Project, 90
human nature, 92–3
Humans 2.0, 97
humility, 29

Icarus, 3
individualism, 139, 159–60, 161
individualistic cultures, 158
individuality, 93
information overload, 147
innovations, 25–6
 acceptance and adoption of, 7, 8, 16
 definition of, 5
 disruptive, 8
 disruptive innovation, 27
 downside of, 4
 risk and, 15–17
 time-tested, 21–2
innovators, 7
integration, 160

Internet, 158, 174, 185, 193–8
  emoticons and, 207–12
  mediated communication and,
    214–19
Internet of Things, 61–6
is-ought problem, 174

Jonathan Livingston Seagull (Bach), 3
junk food, 104–8

laggards, 16
legalized marijuana, 129–33, 148
leisure time, 150
life purpose, 157–62
loss aversion, 15

maker democracy, 50
marijuana, 129–33, 148
Maslow, A.H., 158
mass customization, 49–51
materialism, 138
mean, 175
meaning, 6
meaning-making, 5, 6–7, 16
mediated communication, 207–12,
  214–19
medical devices, 61–2
medicine
  microbiome, 89–94
  pharmaceuticals, 121–7
  preventive gene testing, 81–7
  stem cell therapy, 97–102
memories, 195–6
microbiome, 89–94
Millennials, 133, 141–2, 208
minority groups, 162
mitochrondria, 90
money
  happiness and, 138
  virtual currency, 200–4
mood disorders, 121
moods, 124
moral imagination, 17–18, 20–1
motivation, 123

national happiness goals, 139–40
neurotransmitters, 122, 123
nonviolence, 19
norms, 6, 175

omega-3s, 122
online identity, 193–8
open-mindedness, 130
opportunities, 25–33
optimism, 17
optimum, 174–9
organ donation, 100
originality, 18
outlook, 15–22
over-stimulation, 150
oxytocin, 122, 123, 124–5, 126

parasites, 91–2
parenthood, 142
passivity, 140
personal experiences, 5–6
personal safety, drones and, 70–1
pharmaceuticals, 121–7, 148
play, 18–19
playbook of possibilities, 26
predictions, 20
preventive gene testing, 81–7
printing press, 19
prison population, 130–1
privacy
  Big Data and, 185–6
  drones and, 71–2
  right to be forgotten, 193–8
processed foods, 104–8
Prometheus, 3
prosociality, 18
psychological disorders, 148
puberty, 168
public safety, 131
punishment, 9

recreational drugs, 129–33
responsibility, 10, 91–3
restlessness, 139
right to be forgotten, 193–8
ripple metaphor, 28
risk, 15–17
risk-taking, 126

schools, 149
search engine analytics, 174
security, drones and, 72
self-authenticity, 157–62
selfies, 157

sensemaking, 17
serotonin, 122, 123
serotonin re-uptake inhibitors (SSRIs), 121
sex, 165, 167–8
sexual preference, 165
Sisyphus, 3
skunk works, 25
sleep disorders, 147–8
smart cities, 61
smart homes, 61, 62–3
smartphones, 147
smoking, 111–15
social activism, 7
social connections, 148–9
social institutions, 149–50
social media, 6, 16, 174, 196–7
social responsibility, 133
society, dis-organization of, 49–51
Socrates, 138
solar radiation management, 39
sports, 170
statute of limitations, 196
stem cell therapy, 97–102
systems theory, 5

terminology, 16
testosterone, 122
Theater of the Oppressed, 19
3D printing, 45–51
  in homes, 47
  mass customization and, 49–51
  medical uses of, 48
  throwaway society and, 49
  weapons created by, 48
  workers and, 46–7
thrill seeking, 150
throwaway society, 49
time, 21
tragic hero, 141
transformational imperative, 19
transgender individuals, 169
  see also gender fluidity

uncertainty, 6, 8, 17
Unicode Consortium, 207
unmanned aerial vehicles, 69–74
U.S. Constitution, 19
utensils, 21–2

vaping, 111–15
variability, 175, 179
virtual currency, 200–4
vitamin C, 122
vitamins, 104–8
Vygotsky, L.S., 5

War on Drugs, 129, 130–1
wearable gadgets, 174
workers
  displacement of, 64–5
  impact of 3D printing on, 46–7
work opportunities, 216–17
workplaces, 149
worrying, 15–17

zero, 21